BRITISH OPISTHOBRANCH MOLLUSCS

Plate I. British nudibranchs: top left, *Okenia elegans*; top right, *Thecacera pennigera*; bottom left, *Onchidoris lutescincta*; bottom right, *Coryphella lineata*.

A NEW SERIES

Synopses of the British Fauna
No. 8

BRITISH OPISTHOBRANCH MOLLUSCS

MOLLUSCA: GASTROPODA

Keys and Notes for the Identification of the Species

T. E. THOMPSON

AND

GREGORY H. BROWN

Department of Zoology,
University of Bristol,
Woodland Road,
Bristol, BS8 1UG

1976

Published for
THE LINNEAN SOCIETY OF LONDON
by
ACADEMIC PRESS
LONDON NEW YORK AND SAN FRANCISCO

ACADEMIC PRESS INC. (LONDON) LTD
24–28 Oval Road
London, NW1 7DX

U.S. Edition published by
ACADEMIC PRESS INC.
111 Fifth Avenue
New York, New York 10003

Library of Congress Catalog Card Number: 75–19682
ISBN: 0–12–689350–0

Text set in 9/10 pt. Monotype Times New Roman, printed by photolithography, and bound in Great Britain at The Pitman Press, Bath

Foreword

British Opisthobranch Molluscs is the result of the collaboration of two professional zoologists, one of whom is an invertebrate embryologist and marine biologist while the other is a scientific artist. Both are keen divers. It is hoped that this beautifully illustrated synopsis will interest naturalists in these 'butterflies' of the sea and will inspire aqualung enthusiasts, giving them a greater understanding of the marine world they explore in their underwater travels.

The intertidal naturalist will also find much in this synopsis to interest him, because many of these lovely opisthobranchs may be found in the rock-pools of our coasts.

This synopsis, like its predecessors, is written by specialists for naturalists, both amateur and professional, and is designed to fill the gap between popular comprehensive field-guides and specialist, often rare and valuable, monographs or treatises. This is the first systematic treatment of this group since the Monograph of Alder and Hancock published in 1844–55, and takes account of the many new species and new records described during the last hundred years.

To make these pocket-sized synopses useful as field and laboratory guides, there are spaces left in the text for the owner's notes and the covers are waterproofed, though not to such an extent as to permit total immersion in sea water.

The authors and the Linnean Society are grateful to the University of Bristol for a grant to enable the publication of the colour plate.

Doris M. Kermack
Synopses Editor, Linnean Society

A Synopsis of the British Opisthobranch Molluscs

T. E. THOMPSON

AND

GREGORY H. BROWN

Zoology Department, University of Bristol, England

CONTENTS

Introduction

The Opisthobranchia are a subclass of the gastropod molluscs. They are all hermaphrodite, marine and macroscopic. World-wide there are about 3000 species of opisthobranchs and 5% of these have been recorded from the shallow waters around the British Isles. Of the British species, 30% are infaunal, the remainder epifaunal. It is chiefly the epifaunal type, as exemplified by the aplysiids or the nudibranchs, which has attracted the attention of naturalists for the last century and now, in the nineteen-seventies, has intrigued and captured the interest of the growing body of underwater aqualung enthusiasts.

The opisthobranch subclass has been polyphyletic, numerous evolutionary lines of early prosobranchs having achieved success by abandoning total reliance upon a rigid external shell. In fact, the "sea-slug" facies have arisen many times in more or less independent lines of evolution.

In the species descriptions, information is given upon the known present distribution of the animals concerned. This should not be regarded as complete or comprehensive and information about animals obtained elsewhere would be welcomed by the authors or by the Linnean Society who will pass it on.

General Structure

The opisthobranchs represent an interesting evolutionary offshoot of the Mollusca, in which the precarious balance between the hard and the soft parts has become heavily tilted in favour of the latter. Like other molluscs, they are coelomate invertebrates, and the body is more or less divisible into *head*, *foot* and *visceral mass*. In the more primitive opisthobranchs, such as *Acteon* (Fig. 1D) these three components are clearly visible and a strong coiled external *shell* is present. But in the more advanced forms, for example *Aplysia* (Fig. 1A), *Berthella* (Fig. 1B) and *Philine* (Fig. 1E), the shell has become reduced and internal, covered by the *mantle* (Fig. 1B), and in the highest types such as the sacoglossans (Fig. 1C) and the nudibranchs (Fig. 1F–J) the shell is completely lacking. Where the shell is absent, the body is free to assume all kinds of flamboyant shapes in relation to a wide range of epifaunal life-styles. In these shell-less forms the head, foot and visceral mass have amalgamated in various ways. A glance at Figure 1 will show the reader that adaptive radiation in the opisthobranchs has resulted in a splendid range of morphology.

In world-wide terms, the Opisthobranchia vary in size from the tiny herbivorous sacoglossans which live upon and consume delicate marine algae, one cell at a time, and the sand-dwelling acochlidiaceans, which are sometimes so minute that they can move unharmed between the grains of siliceous and other inorganic material that make up a beach, to the huge *Tochuina tetraquetra* of the Pacific N.W. of the United States, and the *Aplysia* and *Dolabella* species of tropical waters, weighing in air as much as 2 kg. Some may be found only by diving, digging or searching beneath boulders or coral heads; others, like some of the tropical chromodorid nudibranchs, appear to be bold and self-advertising. In British waters we are fortunate in having available for study an especially wide range of opisthobranch species.

The head may be flattened and modified for burrowing as in *Acteon* (Fig. 1D) and *Philine* (Fig. 1E) or may bear tentacles of four distinct kinds. There are the *oral tentacles*, on either side of the mouth (Fig. 1B), the *rhinophoral tentacles* or *rhinophores* on the dorsal side of the head (Fig. 1A), the *propodial tentacles* (Fig. 1D) and the *posterior cephalic tentacles* (Fig. 6A). These are all assumed to be sensory in function. An additional sense organ, the *caruncle*, occurs in *Antiopella* (Fig. 1G), and paired lamellar sense organs occur in certain burrowing forms (e.g. Fig. 10C).

The mantle may be reduced as in *Philine* (Fig. 1E) or expanded so as to cover the dorsum and project all round as a voluminous skirt as in *Archidoris* (Fig. 1F). The *mantle cavity* may be deeply internal (so as to be studied only by dissection) as in *Acteon*, or widely open so that the *gill* projects freely as in *Berthella* (Fig. 1B). In the British sacoglossans and the nudibranchs the mantle cavity with its contained gill has been completely lost. In many nudibranchs new respiratory structures have evolved. These include the *gill circlet* of the dorid nudibranchs (Fig. 1F) and the arborescent lateral gills or cerata of the dendronotaceans (Fig. 1H).

Many of the shell-less opisthobranchs possess other dorsal processes or *cerata*, often brightly marked and containing defensive organs (Fig. 1G, I, J). These

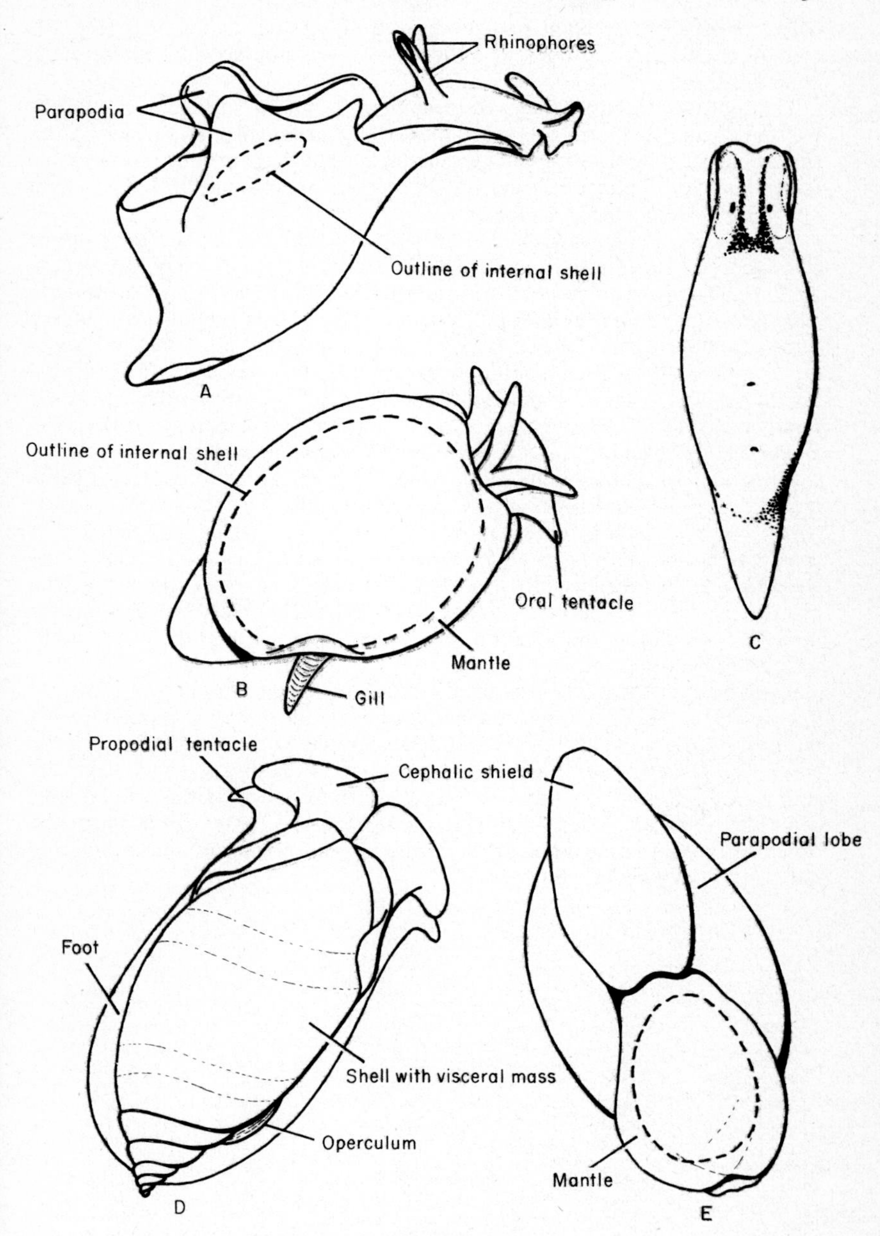

FIG. 1. Range of form in British Opisthobranchia: A, *Aplysia*; B, *Berthella*; C, *Limapontia*; D, *Acteon*; E, *Philine*;

defensive organs may be glandular, or as in the aeolid nudibranchs (Fig. 1J) contain remarkable batteries of sting-cells or nematocysts derived from the cnidarian prey.

The foot always has a flattened creeping sole but in some opisthobranchs has developed conspicuous lateral lobes or *parapodia* (Fig. 1A, E) which are occasionally natatory. Swimming by undulations of these or other parts of the foot is known to occur in certain British species, for example, *Aplysia fasciata* and *Pleurobranchus membranaceus*.

A discussion of the internal anatomy of opisthobranchs is outside the scope of this Synopsis. The interested reader should consult the Ray Society Monograph by T. E. Thompson to be published shortly (*Biology of Opisthobranch Molluscs Volume I*). But new enthusiasts will certainly wish to begin to study the anatomy of opisthobranchs, and Fig. 1K shows a general dissection of *Archidoris pseudoargus*, one of the most common and accessible British sea-slugs. The principal organs visible in a gross dissection from the dorsal side are labelled. In dorid nudibranchs like *Archidoris* there is a high degree of bilateral symmetry. The *mouth* is anterior and the *anus* is posterior. The *buccal mass*, with its complex of extrinsic *muscles*, contains the rasping radula which can be thrust through the mouth and used to scrape up the prey (encrusting sponges in the case of *Archidoris*). The food is lubricated by the secretions of the *salivary glands* then passed along the *oesophagus* to the capacious *stomach*. Digestion begins in the stomach and the products are absorbed in the *digestive gland*. (In aeolids the digestive gland enters the cerata as a series of diverticula.) Wastes (principally the unwanted spicules of the sponges) are passed along the *intestine* to the anus, situated in the centre of the *gill circlet*.

Close beneath the gill circlet various details of the circulatory and excretory systems can be discerned. The muscular *heart* lies in the *pericardial sac*. The heart receives venous blood at the rear and drives arterial blood forward through the *aorta*. The *kidney sac* branches over the surface of the digestive gland and the *reno-pericardial canal* connects the cavities of the kidney and the pericardium.

On the right-hand side of the body the *reproductive organs* can be seen; the organs of both sexes are present in each adult individual and unravelling the individual components is difficult.

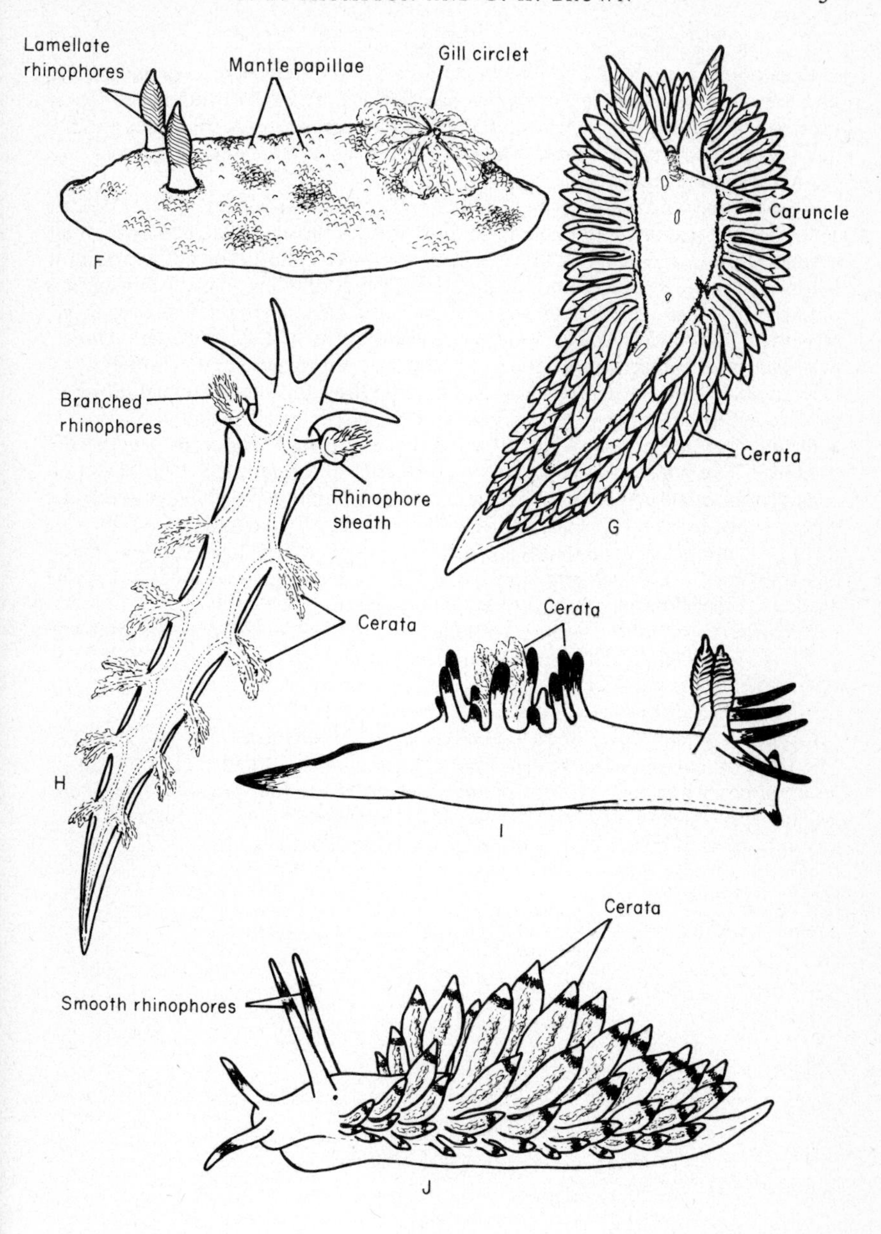

FIG. 1 (contd.) F. *Archidoris*; G. *Antiopella*; H, *Tritonia*; I. *Ancula*; J. *Eubranchus*.

Biology

Opisthobranchs are all free-living animals with well-defined dietary preferences and these are indicated in the species descriptions. The burrowing forms ingest bivalves and tubicolous polychaetes, the sacoglossans are all herbivorous and the nudibranchs are carnivorous. They are all marine, but several species are remarkably euryhaline.

All opisthobranchs are hermaphrodite and copulation is normally reciprocal. The spawn is usually jelly-like and in most species the eggs hatch as swimming shelled *veliger larvae* (Fig. 1L) which spend a relatively short time in the plankton before settling. Settlement and metamorphosis will only occur if certain specific requirements are met. These requirements are often extremely precise, e.g., settlement may occur solely upon some component of the adult diet. Direct, non-pelagic, development occurs in some species, especially in polar waters.

In those British species which have received close study the maximal life-span has proved to be one year or less. Some of these species have annual life cycles, with one breeding period, while others may pass through numerous generations in a year. The purely annual species tend to feed on organisms which have one conspicuous quality in common: their extreme abundance and stability in certain types of locality at all times of the year. Those species which are known to pass through a number of generations in a year are all species which feed upon more or less transitory prey, such as hydroids, which may spring up seasonally on submarine surfaces and be cropped to extinction within a few days.

Records of predation upon opisthobranchs are few. Haddock and cod are known to take *Philine* and other bullomorphs but tests in laboratory aquaria have established that most sea-slugs are rejected as food by a wide variety of carnivorous teleost fishes.

Copepod parasites are often found associated with the British nudibranchs. The tiny ectoparasitic *Lichomolgus agilis* darts over the dorsum of many dorid nudibranchs while the presence of the degenerate internal parasite *Splanchnotrophus* spp. may be detected by the white, sausage-like, twin egg-sacs thrust through the skin of the host, a dorid or aeolid nudibranch.

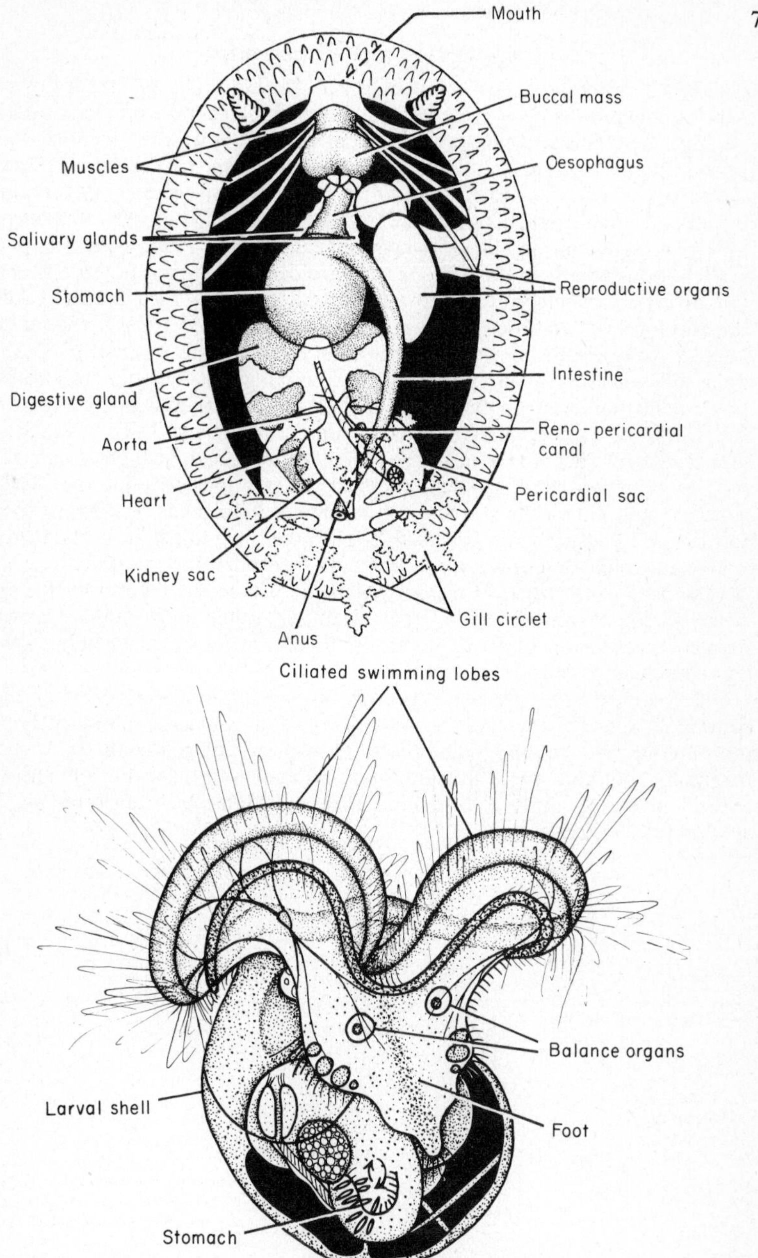

FIG. 1. (contd.) K, dorsal dissection of *Archidoris*, with much of the mantle cut away; L, veliger larva of *Archidoris*.

Collection and Preservation

The shelled bullomorphs are best sought with a spade and a sieve near low water mark on a good spring tide; the rarer sublittoral species may only be taken with a dredge. Aplysiids and sacoglossans occur in shallow water or in pools (sometimes on saltmarshes) associated with species of algae and they are often extremely well camouflaged. The larger nudibranchs are among the easiest opisthobranchs to find and often occur between tide marks on rocky shores. The smaller nudibranchs present considerable problems and can often be found only after patient and laborious search through dredged hydroids, sponges and polyzoans. Aqualung divers are privileged to see the sublittoral species in their natural habitats. The best localities are often the most hazardous and inaccessible. Exposed headlands or rocky islands in areas of rapid tidal currents with water of high salinity are usually the most favourable. Spawn coils often give a clue to the whereabouts of the adult nudibranchs, but many species are conspicuous anyway and such detective work is unnecessary. Obviously no responsible naturalist will take more animals than are needed for his immediate purpose and, in fact, great satisfaction may be obtained from simple ecological observations without the need for capturing and killing the subjects. With difficult individuals or obscure species, however, it is certainly not practicable to attempt identification in the field and it is usually found that opisthobranchs can be transported to the laboratory bench satisfactorily in sea water in a stout polythene bag. In the laboratory the specimens should be transferred to cool clean sea water and not over-crowded. Remember that some of the carnivorous species are voracious when hungry and may attack more delicate specimens.

Observations from life are especially important in these soft and highly contractile animals. Preservation should only follow careful noting of habitat, morphology and colours. Narcotisation may be effected with 7% aqueous magnesium chloride mixed in equal parts with sea water, and the animals when relaxed preserved in 10% formalin. They may later be transferred to 70% alcohol if this is desired.

Classification

CLASS GASTROPODA

Subclass OPISTHOBRANCHIA

Order BULLOMORPHA

Family Acteonidae
Acteon tornatilis (L., 1758)

Family Diaphanidae
Diaphana minuta Brown, 1827
Colpodaspis pusilla M. Sars, 1870

Family Retusidae
Retusa obtusa (Montagu, 1803)
Retusa truncatula (Bruguière, 1792)
Retusa umbilicata (Montagu, 1803)
Rhizorus acuminatus Bruguière, 1792

Family Atyidae
Haminea navicula (da Costa, 1778)
Haminea hydatis (L., 1758)

Family Scaphandridae
Cylichna cylindracea (Pennant, 1777)
Roxania utriculus (Brocchi, 1814)
Scaphander lignarius (L., 1758)
Scaphander punctostriatus (Mighels and Adams, 1841)

Family Akeridae
Akera bullata Müller, 1776

Family Philinidae
Philine aperta (L., 1767)
Philine angulata Jeffreys, 1867
Philine catena (Montagu, 1803)
Philine denticulata (Adams, 1800)
Philine pruinosa (Clark, 1827)
Philine punctata (Adams, 1800)
Philine quadrata (Wood, 1839)
Philine scabra (Müller, 1776)

Family Philinoglossidae
Philinoglossa helgolandica Hertling, 1932
Philinoglossa remanei Marcus and Marcus, 1958

Family Runcinidae
Runcina coronata (Quatrefages, 1844)

Order APLYSIOMORPHA

Family Aplysiidae

Aplysia punctata Cuvier, 1803
Aplysia depilans Gmelin, 1791
Aplysia fasciata Poiret, 1789

Order PLEUROBRANCHOMORPHA

Family Pleurobranchidae

Pleurobranchus membranaceus (Montagu, 1815)
Berthella plumula (Montagu, 1803)
Berthellina citrina (Rüppell and Leuckart, 1828)

Order ACOCHLIDIACEA

Family Hedylopsidae

Hedylopsis brambelli Swedmark, 1968
Hedylopsis spiculifera Kowalevsky, 1901
Hedylopsis suecica Odhner, 1937

Family Microhedylidae

Microhedyle lactea Hertling, 1930

Order SACOGLOSSA

Family Stiligeridae

Hermaea bifida (Montagu, 1815)
Hermaea dendritica (Alder and Hancock, 1843)
Hermaea variopicta (Costa, 1869)
Stiliger bellulus (Orbigny, 1837)
Alderia modesta (Lovén, 1844)

Family Elysiidae

Elysia viridis (Montagu, 1808)

Family Limapontiidae

Limapontia capitata (Müller, 1774)
Limapontia depressa Alder and Hancock, 1862
Limapontia senestra (Quatrefages, 1844)

Order NUDIBRANCHIA

Suborder DENDRONOTACEA

Family Tritoniidae

Tritonia hombergi Cuvier, 1803
Tritonia plebeia Johnston, 1828
Tritonia lineata Alder and Hancock, 1848
Tritonia odhneri (Tardy, 1963)

Family Lomanotidae
Lomanotus marmoratus (Alder and Hancock, 1845)

Family Dendronotidae
Dendronotus frondosus (Ascanius, 1774)

Family Scyllaeidae
Scyllaea pelagica L., 1758

Family Hancockiidae
Hancockia uncinata (Hesse, 1872)

Family Dotoidae
Doto cinerea Trinchese, 1881
Doto coronata (Gmelin, 1791)
Doto cuspidata Alder and Hancock, 1862
Doto fragilis (Forbes, 1838)
Doto pinnatifida (Montagu, 1804)

Suborder DORIDACEA
Family Goniodorididae
Goniodoris nodosa (Montagu, 1808)
Goniodoris castanea Alder and Hancock, 1845
Okenia elegans (Leuckart, 1828)
Okenia aspersa (Alder and Hancock, 1845)
Okenia leachi (Alder and Hancock, 1854)
Okenia pulchella (Alder and Hancock, 1854)
Ancula cristata (Alder, 1841)
Trapania pallida Kress, 1968
Trapania maculata Haefelfinger, 1960

Family Onchidorididae
Acanthodoris pilosa (Müller, 1789)
Adalaria proxima (Alder and Hancock, 1854)
Adalaria loveni (Alder and Hancock, 1862)
Onchidoris bilamellata (L., 1767)
Onchidoris muricata (Müller, 1776)
Onchidoris depressa (Alder and Hancock, 1842)
Onchidoris inconspicua (Alder and Hancock, 1851)
Onchidoris oblonga (Alder and Hancock, 1845)
Onchidoris sparsa (Alder and Hancock, 1846)
Onchidoris luteocincta (M. Sars, 1870)
Onchidoris pusilla (Alder and Hancock, 1845)

Family Triophidae
Crimora papillata Alder and Hancock, 1862

Family Notodorididae
Aegires punctilucens (Orbigny, 1837)

Family Polyceridae
Polycera quadrilineata (Müller, 1776)
Polycera faeroensis Lemche, 1929
Palio dubia M. Sars, 1829
Greilada elegans Bergh, 1894
Thecacera pennigera (Montagu, 1815)
Thecacera capitata Alder and Hancock, 1854
Thecacera virescens Forbes and Hanley, 1851
Limacia clavigera (Müller, 1776)

Family Cadlinidae
Cadlina laevis (L., 1767)

Family Aldisidae
Aldisa zetlandica (Alder and Hancock, 1854)*
Aporodoris millegrana (Alder and Hancock, 1854)

Family Rostangidae
Rostanga rubra (Risso, 1818)

Family Dorididae
Doris verrucosa L., 1758
Doris maculata Garstang, 1895

Family Archidorididae
Archidoris pseudoargus (Rapp, 1827)
Atagema gibba Pruvot-Fol, 1951

Family Discodorididae
Discodoris planata (Alder and Hancock, 1846)

Family Kentrodorididae
Jorunna tomentosa (Cuvier, 1804)

Suborder ARMINACEA

Family Arminidae
Armina loveni (Bergh, 1860)

Family Antiopellidae
Antiopella cristata (Chiaje, 1841)
Antiopella hyalina (Alder and Hancock, 1854)
Proctonotus mucroniferus (Alder and Hancock, 1844)

Family Heroidae
Hero formosa (Lovén, 1841)

*This species is little known but is placed in this family provisionally.

Suborder AEOLIDACEA

Family Coryphellidae
Coryphella lineata (Lovén, 1846)
Coryphella pedata (Montagu, 1815)
Coryphella verrucosa (M. Sars, 1829)

Family Facelinidae
Facelina auriculata (Müller, 1776)
Facelina annulicornis (Chamisso and Eysenhard, 1821)

Family Favorinidae
Favorinus blianus Lemche and Thompson, 1974
Favorinus branchialis (Rathke, 1806)

Family Aeolidiidae
Aeolidia papillosa (L., 1761)
Aeolidiella alderi (Cocks, 1852)
Aeolidiella glauca (Alder and Hancock, 1845)
Aeolidiella sanguinea (Norman, 1877)

Family Eubranchidae
Eubranchus tricolor Forbes, 1838
Eubranchus farrani (Alder and Hancock, 1844)
Eubranchus pallidus (Alder and Hancock, 1842)
Eubranchus vittatus (Alder and Hancock, 1842)
Eubranchus cingulatus (Alder and Hancock, 1847)
Eubranchus exiguus (Alder and Hancock, 1848)

Family Cumanotidae
Cumanotus beaumonti (Eliot, 1906)

Family Pseudovermidae
Pseudovermis boadeni Salwini-Plawen and Sterrer, 1968

Family Cuthonidae
Precuthona peachi (Alder and Hancock, 1848)
Cuthona nana (Alder and Hancock, 1842)
Catriona aurantia (Alder and Hancock, 1842)
Tenellia pallida (Alder and Hancock, 1842)
Embletonia pulchra (Alder and Hancock, 1851)
Tergipes tergipes (Forskål, 1775)
Trinchesia caerulea (Montagu, 1804)
Trinchesia amoena (Alder and Hancock, 1845)
Trinchesia concinna (Alder and Hancock, 1843)
Trinchesia viridis (Forbes, 1840)
Trinchesia foliata (Forbes and Goodsir, 1839)

Family Fionidae
Fiona pinnata (Eschscholtz, 1831)

Family Calmidae
Calma glaucoides (Alder and Hancock, 1854)

Key to the Families of British Opisthobranchia

Note: The key is based upon external features and applies only to families as represented by British species of Bullomorpha, Aplysiomorpha, Pleurobranchomorpha, Acochlidiacea, Sacoglossa and Nudibranchia (i.e. excluding pteropods and pyramidellids). The pyramidellids are included in another Synopsis in this series, Graham: *British Prosobranch and other Operculate Gastropod Molluscs* (New Series) No. 2.

1. External shell present **2**
Shell internal (covered by the mantle) or absent **7**

2. Operculum present (Fig. 1D) . . . . . . Family Acteonidae (p. 17)
Operculum absent **3**

3. Rear of foot forked . . . . Family Diaphanidae (part) (p. 18)
Rear of foot entire **4**

4. Cephalic shield bearing postero-lateral tentacles Family Retusidae (p. 20)
Cephalic shield without conspicuous postero-lateral tentacles . . . **5**

5. External shell stout and strong . . . . . Family Scaphandridae (p. 26)
External shell inflated but frail **6**

6. Parapodial lobes natatory . . . . Family Akeridae (p. 30)
Parapodial lobes well developed but not natatory . Family Atyidae (p. 24)

7. Shell internal **8**
Shell absent **11**

8. Body shape quadripartite (Fig. 1E) . . . . Family Philinidae (p. 31)
Body shape otherwise **9**

9. Internal shell inflated, spiral . . . . Family Diaphanidae (part) (p. 18)
Internal shell widely open, cap-like (Fig. 1B) **10**

10. Sea hares (Fig. 1A) . . . . Family Aplysiidae (p. 38)
Side-gilled flattened slugs (Fig. 1B) . . Family Pleurobranchidae (p. 41)

11. Burrowing sea-slugs in shell-gravel **12**
Epifaunal sea-slugs **15**

12. Head lacking tentacles **13**
Head tentacles conspicuous **14**

13. Vestigial cerata (paired dorsal papillae) present
Family Pseudovermidae (p. 174)
Body smooth, lacking such papillae . . . Family Philinoglossidae (p. 36)

14. Oral tentacles much larger than rhinophores . Family Hedylopsidae (p. 46)
Oral tentacles slightly larger than rhinophores
Family Microhedylidae (p. 48)

15. Postero-lateral gill plume present, under mantle rim
Family Runcinidae (p. 37)
Postero-lateral gill absent **16**

Systematic Part

In the description of many of the species there is given, below the name of the species, the name under which it was first described and, in some cases, a few of the synonyms. The keys are based upon external features only, but in some cases it will be necessary to open up the mantle to ascertain if an internal shell is present.

Family ACTEONIDAE

Genus ACTEON (Montfort, 1810)

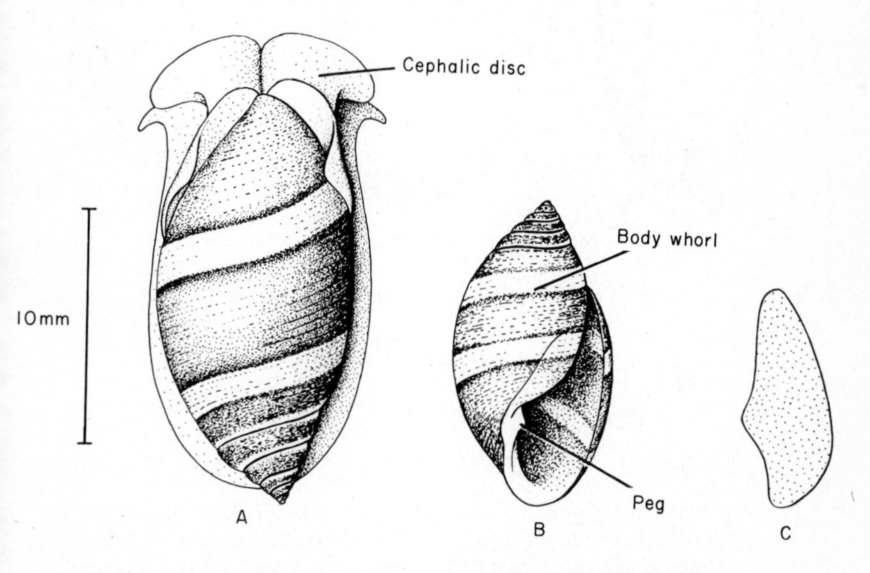

FIG. 2. *Acteon tornatilis:* A, whole animal; B, shell; C, operculum.

Acteon tornatilis (L., 1758) (Fig. 2A–C)

Shell solid, opaque, glossy, light pink, with 1–3 white bands on the body whorl and one on each of the other major whorls; up to 8 whorls in all. A conspicuous tooth or peg is present within the mouth of the shell (Fig. 2B). Operculum (Fig. 2C) amber-coloured, frail, flattened.

The overall length of the creamy white body may reach 30 mm. The cephalic disc is flattened and functions as a plough when the animal burrows through the sand in which it dwells, from LWST down to 250 m. When exposed to the air, inter-tidal individuals come to the surface and then may easily be captured. This species is entirely carnivorous and feeds upon infaunal polychaetes.

It is generally distributed in sheltered sandy bays around the British Isles; further distribution from Iceland to the Mediterranean.

Family DIAPHANIDAE

Genus DIAPHANA Brown, 1827

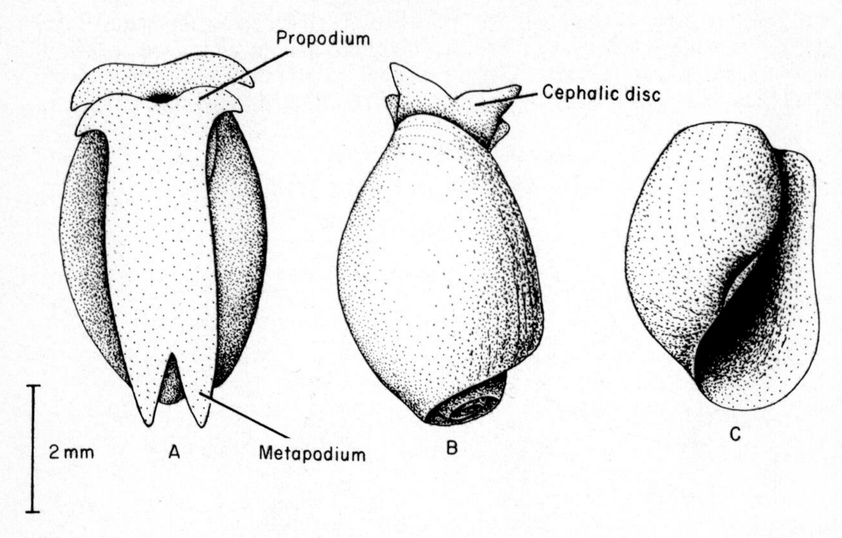

FIG. 3. *Diaphana minuta:* A, ventral view; B, dorsal view; C, shell (after Sars (1878) and Forbes and Hanley (1848–53)).

Diaphana minuta Brown, 1827 (Fig. 3A–C)

Shell fragile, delicate, translucent white; up to 5 whorls in all. Spire usually flattened.

The body may reach 8 mm in overall length and is white with a tinge of amber. The cephalic disc is produced laterally to form flattened tentacular processes. The propodium is dilated antero-laterally so as to form a pair of wing-like tentacular processes. The metapodium is forked posteriorly. The diet is unknown.

The animal burrows in sand, from LWST down to 350 m. This species is generally distributed all round the British Isles, and may be common in some localities; further distribution from Greenland to New England and from Iceland and the Arctic Sea to the Mediterranean and the Canary Islands.

Genus COLPODASPIS M. Sars, 1870

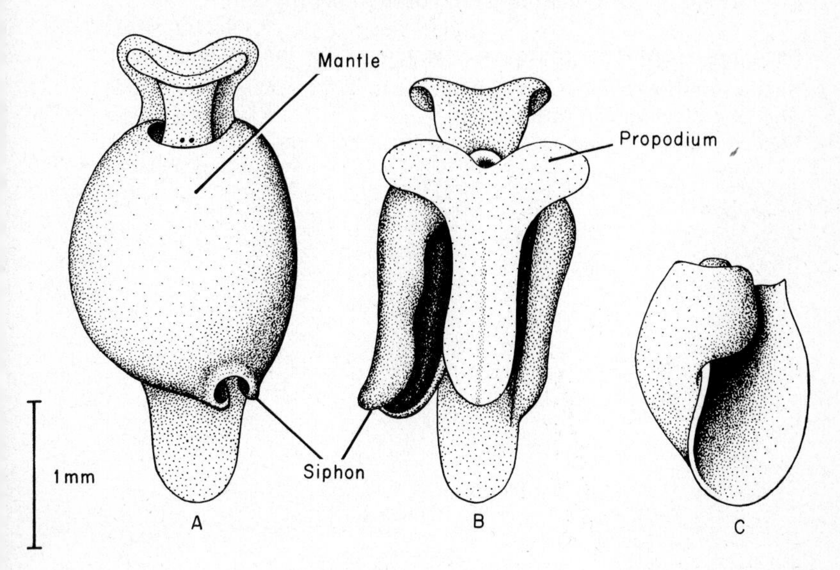

FIG. 4. *Colpodaspis pusilla:* A, dorsal view; B, ventral view; C, shell (after Garstang (1894)).

Colpodaspis pusilla M. Sars, 1870 (Fig. 4A–C)

Shell internal, covered by the mantle, delicate, translucent white, with a somewhat flattened spire and a sharply flaring outer lip; up to 3 whorls in all.

The body measures up to 3 mm in length, white with opaque white speckling. The propodium is dilated antero-laterally to form a pair of wing-like processes. The mantle is smooth and is produced posteriorly on the right side so as to form a conspicuous siphon.

There are only two British records, from rough ground in 30 m of water near Plymouth and from shallow water around Lundy. Elsewhere, it is known from depths down to 160 m off the Norwegian coast.

Family RETUSIDAE

1. Shell-length up to 10 mm; aperture shorter than the spire *Retusa obtusa* (p. 21)
Shell-length up to 5 mm; aperture as long as, or longer, than the spire **2**

2. Shell smoothly rounded, broadest in the middle *Retusa umbilicata* (p. 23)
Shell more cylindrical in shape . . . . . . *Retusa truncatula* (p. 22)
Shell acuminate (elongated and tapering at both ends) *Rhizorus acuminatus* (p. 23)

Genus RETUSA Brown, 1827

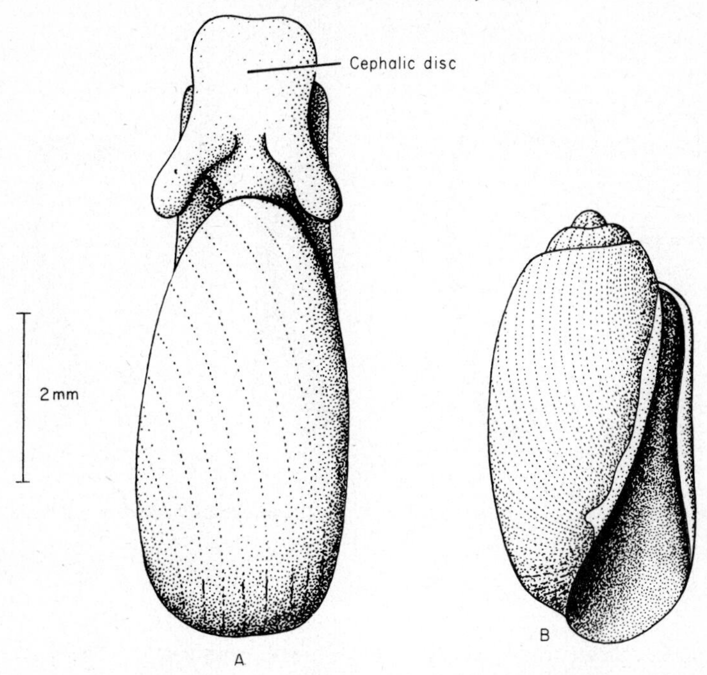

FIG. 5. *Retusa obtusa:* A, dorsal view; B. shell.

Retusa obtusa (Montagu, 1803) (Fig. 5A and B)

Shell fragile, delicate, translucent white, with a spire which may be flattened (Fig. 5A) or slightly produced (Fig. 5B).

The body measures up to 15 mm in length, whitish. The cephalic disc is produced postero-laterally to form ear-shaped tentacular processes. Foot short and squat. Conspicuous epizoitic stalked ciliate protozoans are commonly present on live shells. This species burrows shallowly in mud or muddy sand and feeds upon the prosobranch *Hydrobia ulvae* (Pennant) which is swallowed whole.

This species is generally distributed around the British Isles and, in favourable localities, like the muddy beach at Weston-super-Mare, is extremely abundant; further distribution from Greenland and Iceland to Scandinavia, from Nova Scotia, and from the Aleutian Islands, down to 300 m.

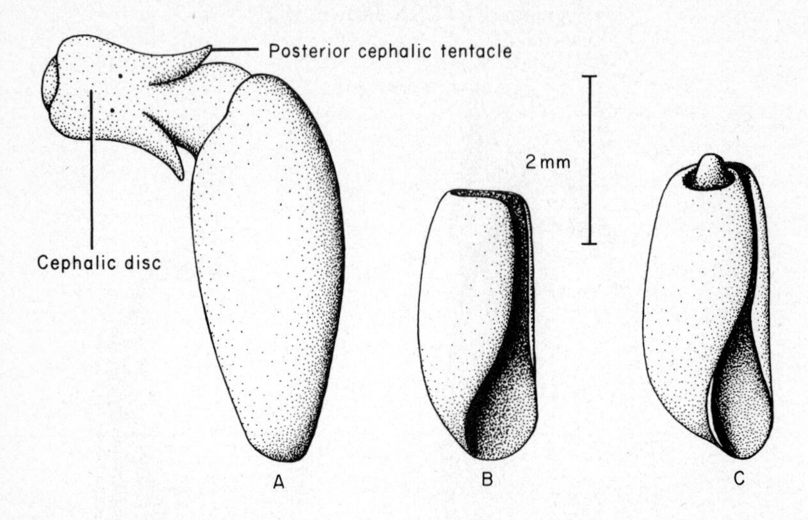

FIG. 6. *Retusa truncatula:* A, dorsal view; B, C, shell-varieties (after Meyer and Möbius (1865)).

Retusa truncatula (Bruguière, 1792) (Fig. 6A–C)

Shell fragile, delicate translucent white or yellowish, with a spire which may be flattened (Fig. 6B) or slightly produced (Fig. 6C).

The white body may reach 7 mm in length. The cephalic disc is produced postero-laterally to form pointed tentacular processes. Foot short and squat. It feeds upon small molluscs and upon foraminiferans.

This species lives in sand and mud all around the coasts of Britain, from the lower shore down to 50 m or more. Further distribution from Norway and the Baltic Sea to the Mediterranean Sea and the Canary Islands, down to 200 m.

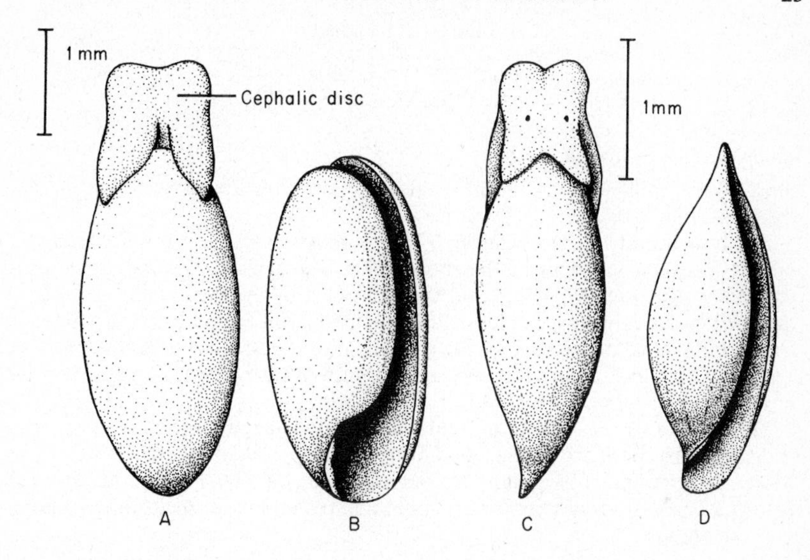

FIG. 7. *Retusa umbilicata:* A, dorsal view; B, shell; *Rhizorus acuminatus:* C, dorsal view; D, shell (after Sars (1878) and Pilsbry (1893)).

Retusa umbilicata (Montagu, 1803) (Fig. 7A and B)

Shell fragile, delicate, translucent white, with a non-protruding spire (Fig. 7B).

The white body may reach 6 mm in length. Cephalic disc produced postero-laterally to form somewhat pointed (less sharply so than in Fig. 6A *R. truncatula*) tentacular processes. Foot short and squat. The diet is unknown.

This species occurs all around the British Isles, usually in muddy sand; further distribution from Norway and the Baltic Sea to the Mediterranean Sea, Madeira and the Cape of Good Hope, together with reports of what is probably the same species from the New England coast. Depth records range down to 2000 m.

Genus RHIZORUS Montfort, 1810

Rhizorus acuminatus Bruguière, 1792 (Fig. 7C and D)

Shell fragile, delicate, translucent white, slender, especially narrowed (acuminated) at the two extremities (Fig. 7D).

The pale body measures up to 6 mm in length. The cephalic disc is produced postero-laterally to form flattened, ear-shaped (not sharply pointed) tentacular processes. The diet is unknown.

It lives sublittorally in muddy sand all around the British Isles. Further distribution from Norway to the Mediterranean and Angola, down to 800 m.

Family ATYIDAE

Genus HAMINEA Turton, 1830

Haminea navicula (da Costa, 1778) (Fig. 8A–C)

Bulla navicula da Costa, 1778

Shell fragile, inflated, translucent white or yellow with the aperture longer than the spire (Fig. 8B).

The brown body measures up to 7 mm in length with substantial parapodial lobes concealing the sides of the shell in life. The cephalic disc bears flattened tentacular processes which conceal the front part of the shell, while the rear of the shell is overlain by a posterior lobe of the mantle. A pectinate sense organ lies on either side of the edge of the cephalic disc; these sense organs are termed organs of Hancock and in this species each such organ consists of up to 20 pairs of lamellae. The diet is unknown.

This species occurs in littoral mud and in muddy sand, especially on *Zostera* beds. Such habitats are rare around British coasts, unlike the coasts of France where this species is abundant in, for example, the Bassin d'Arcachon. The British records are fifty years or more old and relate only to the southern shores of these islands.

Haminea hydatis (L., 1758) (Fig. 8D)

Bulla hydatis L., 1758

Shell and body similar to *H. navicula*, but adults of *H. hydatis* never surpass an overall body length of 30 mm and the organ of Hancock on each side of the body has up to 12 pairs of lamellae. Internally, the prostate gland can be seen to consist of 3 distinct regions (Fig. 8D) whereas this organ in *H. navicula* consists of only 2 regions (Fig. 8C). There are also clear-cut differences in the mode of development of the eggs and in their methods of locomotion of the two species which remove doubts about the distinctness of the two species. *H. hydatis* is known to swim and this certainly does not occur in *H. navicula*.

The diet can include small bivalve molluscs although some authorities consider these opisthobranchs to be principally herbivorous.

H. hydatis occurs in muddy sand or sand, from the lower shore to unknown depths, on southern coasts of the British Isles. Further distribution (assuming early identifications, based upon shell features alone, were correct) from Atlantic France to the Mediterranean Sea and the Canary Islands, Ascension I. and St. Helena.

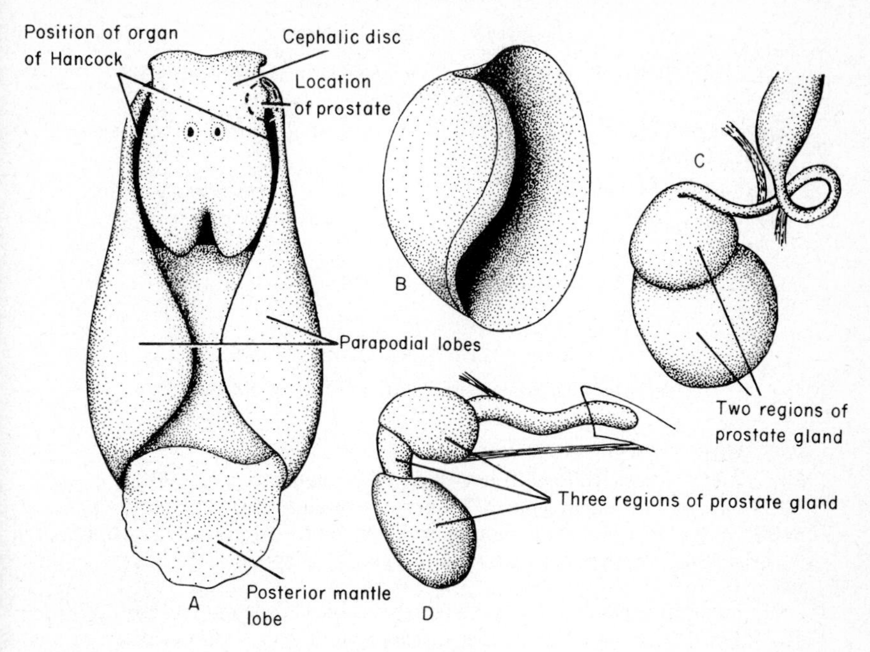

FIG. 8. *Haminea navicula:* A, dorsal view; B, shell; C, prostate gland; *Haminea hydatis:* D, prostate gland.

Family SCAPHANDRIDAE

1. Shell aperture considerably longer than the spire . *Roxania utriculus* (p. 26)
Shell aperture the same length as the spire **2**

2. Shell cylindrical . . . . *Cylichna cylindracea* (p. 26)
Shell inflated **3**

3. Shell a tear-drop in shape . . . . . . *Scaphander lignarius* (p. 28)
Shell more oval . . . . *Scaphander punctostriatus* (p. 28)

Genus CYLICHNA Lovén, 1846

Cylichna cylindracea (Pennant, 1777) (Fig. 9A and B)

Bulla cylindracea Pennant, 1777

Shell solid, opaque, white with brownish areas, elongated (Fig. 9B).

The pale body measures up to 20 mm in overall length. The cephalic disc is produced posteriorly into smoothly rounded tentacular processes. A reddish defensive fluid may be emitted if attacked. This species feeds upon minute shelled protozoans.

It is found in sublittoral fine sand (rarely on the shore) and has been recorded all around the British Isles; further distribution down to 1500 m, from Iceland to the Atlantic and Mediterranean coasts of France, the Azores and the Canary Islands.

Genus ROXANIA Gray, 1847

Roxania utriculus (Brocchi, 1814) (Fig. 9C and D)

Bulla utriculus Brocchi, 1814

Shell solid, translucent, white to pale yellow with brownish areas (Fig. 9D). The shell mouth is considerably longer than the spire.

The overall length of the grey-brown body can reach 18 mm. The rear extremities of the cephalic disc are produced to form pointed tentacular processes. Conspicuous parapodial lobes and a posterior mantle lobe conceal much of the shell. The natural diet is unknown but it is found in sublittoral muddy sand. It is said to swim with ease if disturbed, but this has yet to be confirmed.

R. utriculus occurs all round the British Isles and old reports imply that it may be an important source of food for haddock. Further distribution from north-west Europe to the Mediterranean Sea and the Canary Islands, down to 1500 m in depth.

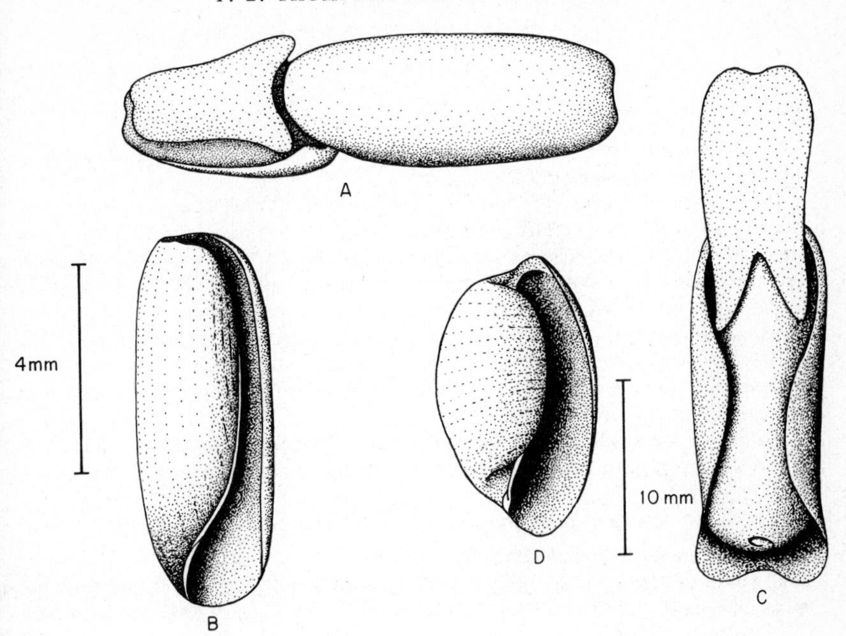

FIG. 9. *Cylichna cylindracea:* A, lateral view; B, shell; *Roxania utriculus:* C, dorsal view; D, shell (after Lemche (1948) and Pilsbry (1893)).

Genus SCAPHANDER Montfort, 1810

Scaphander lignarius (L., 1758) (Fig. 10A–C)

Bulla lignaria L., 1758

Shell solid, opaque, glossy, green, yellow or brown (Fig. 10B). The shell mouth is slightly longer than the spire.

The white or orange body may reach 60 mm in overall length. Large fleshy lobes of the foot and mantle are present but do not conceal the shell and the organ of Hancock can be seen nestling under the cephalic disc (Fig. 10C). A viscous yellow fluid may be discharged on disturbance. This species is very common in sublittoral sand, burrowing beneath the surface in search of its prey, chiefly small worms, bivalve molluscs and other burrowing invertebrates. It has in the past been taken most usually in the dredge but aqualung divers may in present times have encountered it without thinking it noteworthy. It is known to be a source of food for haddock.

S. lignarius occurs all round the British Isles; further distribution from Iceland to the Mediterranean Sea and the Canary Islands, down to 700 m in depth.

Scaphander punctostriatus (Mighels and Adams, 1841) (Fig. 10D)

Bulla punctostriata Mighels and Adams, 1841

Shell solid, opaque, glossy, pallid, with the shell mouth slightly longer than the spire. The shell shape is more smoothly ovoid than that of *S. lignarius* (compare Figs 10B and 10D).

The apearance in life and the natural history of this deep-water species are not well documented and more records and information are urgently needed. In size *S. punctostriatus* is slightly smaller than the common *S. lignarius*.

S. punctostriatus occurs only off our northern coasts; elsewhere it has been recorded from Greenland and the Iceland to Newfoundland and Massachusetts. Records exist also for the Mediterranean, Canary Islands, West Indies, India and Australia but these need confirmation.

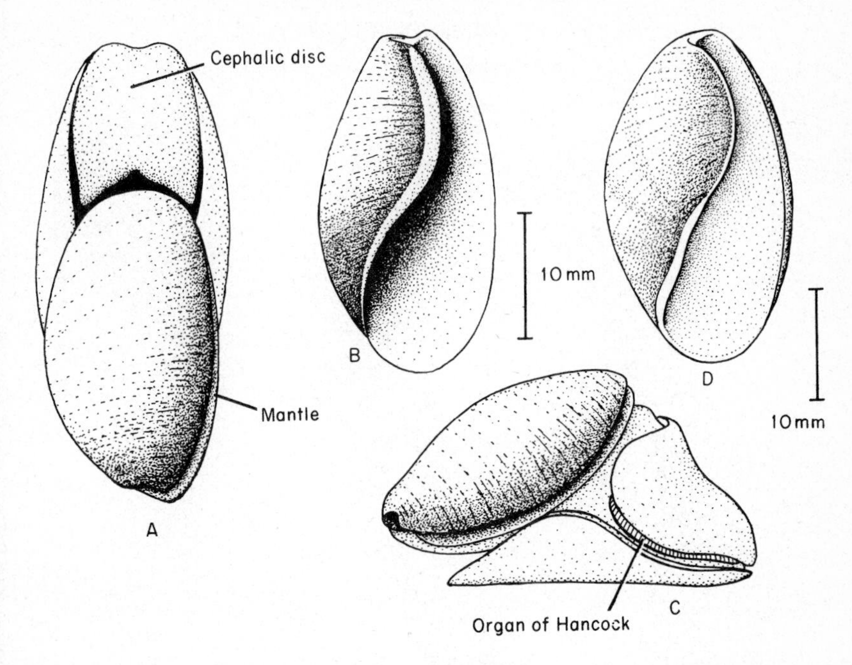

FIG. 10. *Scaphander lignarius:* A, dorsal view; B, shell; C, lateral view; *Scaphander punctostriatus:* D, shell.

Family AKERIDAE

Genus AKERA Müller, 1776

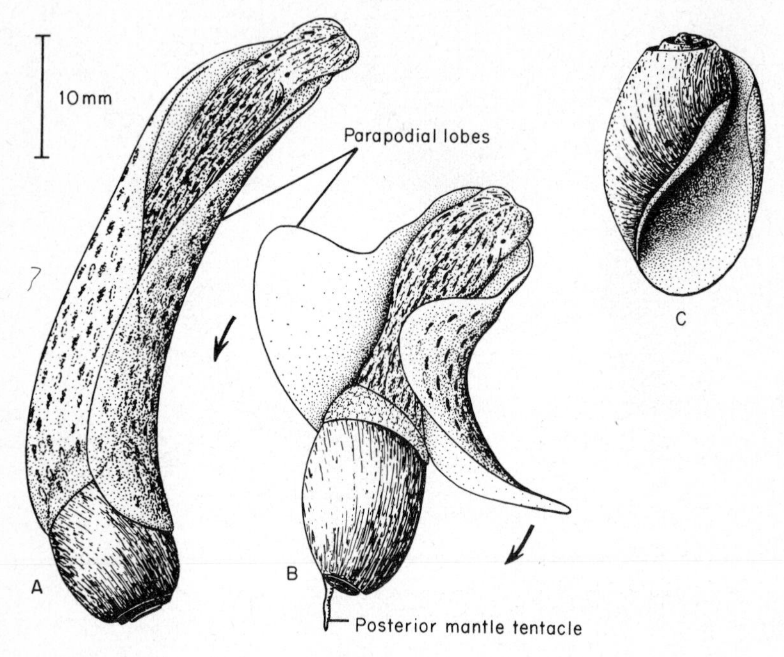

FIG. 11. *Akera bullata:* A, dorsal view; B, swimming individual; C, shell.

Akera bullata Müller, 1776 (Fig. 11A–C)

Shell swollen, fragile, glossy, white to pale brown; up to 6 whorls in all, with the shell mouth slightly shorter than the spire (Fig. 11C).

The body may reach 60 mm in overall length, varying in ground colour from pale grey to orange, with many small white and dark spots, and often streaked in front with blotchy lines or purplish brown. Large parapodial lobes are present, covering the shell in part when at rest. A purple fluid may be secreted if abruptly disturbed; some races appear to lack this capability. Similarly variable is the presence or absence of a posterior mantle tentacle (Fig. 11B). Swimming is sporadic and is brought about by graceful synchronous movements of the parapodia (Fig. 11B). This herbivorous species is found on and in soft fine mud, such as occurs in the sheltered bays where *Zostera* was once abundant. It is itself known to be eaten by flounders and dabs.

Akera bullata occurs on the southern coasts of England, Scotland, Wales and Ireland; further distribution from Norway to the Mediterranean Sea, down to 370 m in depth.

Family PHILINIDAE

1. Body-length up to 70 mm, white in colour . . . . *Philine aperta* (p. 32)
Body-length up to 17 mm, often brown in colour **2**

2. Posterior border of shell forms a single large denticle or spine
Philine angulata, *P. denticulata* (p. 32 and 33)
Posterior border of shell rounded
Philine pruinosa, *P. punctata*, *P. quadrata*, *P. scabra*, *P. catena* (p. 33–34)

Genus PHILINE Ascanius, 1772

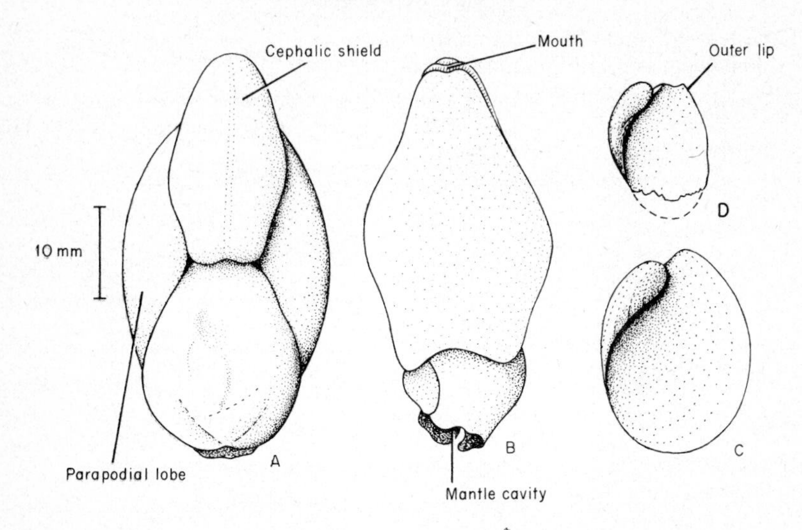

FIG. 12. *Philine aperta:* A, dorsal view; B, ventral view; C, shell; *Philine angulata:* D, shell (slightly damaged) (after Lemche (1948)).

Philine aperta (L., 1767) (Fig. 12A–C)

Bulla aperta L., 1767

B. quadripartita Ascanius, 1772

Shell internal (covered by the mantle), oval, fragile, transparent, glossy, pallid; the aperture very wide and open, longer than the spire (Fig. 12C).

The body may reach 70 mm in overall length, white to pale yellow in colour, approximately quadrangular in shape. The mantle cavity is widely open posteriorly. The skin can secrete sulphuric acid as a defensive reaction. This extremely common species occurs in offshore sand, through which it burrows in search of its prey, consisting of small gastropods, bivalves and polychaete worms. It may itself be taken by fish such as the haddock and some flatfish.

Philine aperta occurs sublittorally all round the British Isles; further distribution from Norway to the Mediterranean Sea, South Africa, Ceylon, and the Philippines, to 500 m in depth.

Philine angulata Jeffreys, 1867 (Fig. 12D)

This is a dubious species, and its appearance in life is unknown. Its most clear distinguishing feature is to be found in the tiny shells (up to 3 mm long) which have the posterior margin of the outer lip produced to form a sharp point (Fig. 12D).

Further records are urgently needed.

Outside the British Isles it is known from Norway to the Mediterranean Sea and from north-east America to 160 m in depth.

Philine catena (Montagu, 1803) (Fig. 13D–F)

Bulla catena Montagu, 1803

Shell internal (covered by the mantle), oval, fragile, semi-transparent, glossy, whitish; the aperture very wide and open, longer than the spire (Fig. 13E). Minute sculpture consisting of spiral rows of regular indentations (Fig. 13F).

The body may reach 7 mm in overall length, cream or pale brown in colour, with tiny specks of dark brown, orange and white; elongated shape (Fig. 13D). This uncommon species may occur under boulders on British shores, but it is less rare sublittorally. The natural diet is not known, but this philinid is known to be itself eaten by flatfish.

Philine catena occurs all round the British Isles; further distribution from Lofoten to the Mediterranean Sea and the Canary Islands, to 2000 m in depth.

Philine denticulata (Adams, 1800) (Fig. 13G, H)

Bulla denticulata Adams, 1800

Philine sinuata Stimpson, 1850

Philine nitida Jeffreys, 1867

Shell thin, fragile, white, covered by the mantle to a considerable extent, but the upper part is naked even in the adult. The aperture is approximately as long as the spire; the posterolateral border of the outer lip is flared and acutely pointed (Fig. 13G).

The body does not exceed 4 mm in overall length, pale brown in colour, with darker patches. The rear of the mantle is dilated (Fig. 13H). The natural diet is uncertain.

This uncommon species occurs sublittorally all round the British Isles and there is a North Wales record from intertidal mud. Elsewhere it has been found from Norway to the Mediterranean Sea, usually in shallow water.

Philine pruinosa (Clark, 1827) (Fig. 13I–K)

Laona pruinosa (Clark, 1827)

Bullaea pruinosa Clark, 1827

Shell internal (covered by the mantle), strong, translucent white; the aperture wide and open, approximately as long as the spire (Fig. 13I). Minute sculpture consisting of rows of raised dots, in some areas running together to form lines (Fig. 13J).

The body may be up to 17 mm long, white, with an oval shape and rather smaller parapodial lobes than are usual in this genus (Fig. 13K).

This uncommon species occurs all round the British Isles, sublittorally, but nothing is known of its ecology so information about it would be welcome. Elsewhere it has been reported from Norway to the Mediterranean Sea, to 400 m in depth.

Philine punctata (Adams, 1800) (Fig. 13L–N)

Bulla punctata Adams, 1800

Bullaea alata Forbes, 1844

Shell internal (covered by the mantle), oval, fragile, white; the aperture very wide, approximately as long as the spire (Fig. 13M). Minute sculpture consisting of spiral rows of oval dots which occasionally fuse in some areas of the shell (Fig. 13N).

The body may reach 5 mm in overall length, light yellow in colour, with red-brown specks; oval shape with gaping mantle cavity.

This uncommon philinid may be found in lower shore pools and in sublittoral silty sand, all around British coasts, but very little is known about its ecology. Further distribution from Greenland, Shetland and Norway to the Mediterranean Sea, to 240 m in depth.

Philine quadrata (Wood, 1839) (Fig. 13O–Q)

Bulla quadrata Wood, 1839

Philine scutulum Lovén, 1846

Shell internal (covered by the mantle), oval to round, thin, fragile, translucent white; the aperture very wide and open, approximately as long as the spire (Fig. 13P). Minute sculpture consisting of spiral rows of oval dots, linked together in a chain-like manner (Fig. 13Q).

The overall length of the body is up to 16 mm, whitish and oval in shape (Fig. 13O).

This is another uncommon species, occurring sublittorally in sand all round the British Isles. Very little is known of the ecology of *P. quadrata*. Further distribution from Greenland, Iceland and the White Sea to the Mediterranean Sea, Azores and St. Helena. It is also recorded from the New England coast of north America. The maximal depth recorded as 2150 m.

Philine scabra (Müller, 1776) (Fig. 13A–C)

Bulla scabra Müller, 1776

Shell internal (covered by the mantle), oval, more cylindrical than other philinids, thin, fragile, translucent white; the aperture wide, shorter than the spire (Fig. 13B). Minute sculpture consisting of spiral rows of oval dots giving a chain-like appearance. These chains project in a saw-tooth fashion along the anterior part of the shell's outer lip (Fig. 13B, C).

The body may reach 20 mm in overall length, white, sometimes with dark specks; elongated and cylindrical in shape (Fig. 13A). Its natural diet is unknown but it is known that *P. scabra* may be eaten by flatfish.

This distinctive species occurs in sand all round the British Isles, usually sublittorally. Further distribution from Iceland and Norway to the Mediterranean Sea, West Africa and Madeira.

FIG. 13. *Philine scabra:* A, dorsal view; B, shell; C, outer lip magnified; *Philine catena:* D, dorsal view; E, shell; F, sculpture magnified; *Philine denticulata:* G, shell; H, dorsal view; *Philine pruinosa:* I, shell; J, sculpture magnified; K, dorsal view; *Philine punctata:* L, dorsal view; M, shell; N, sculpture magnified; *Philine quadrata:* O, dorsal view; P, shell; Q, sculpture magnified (after Forbes and Hanley (1848–53), Sars (1878) and Lemche (1948)).

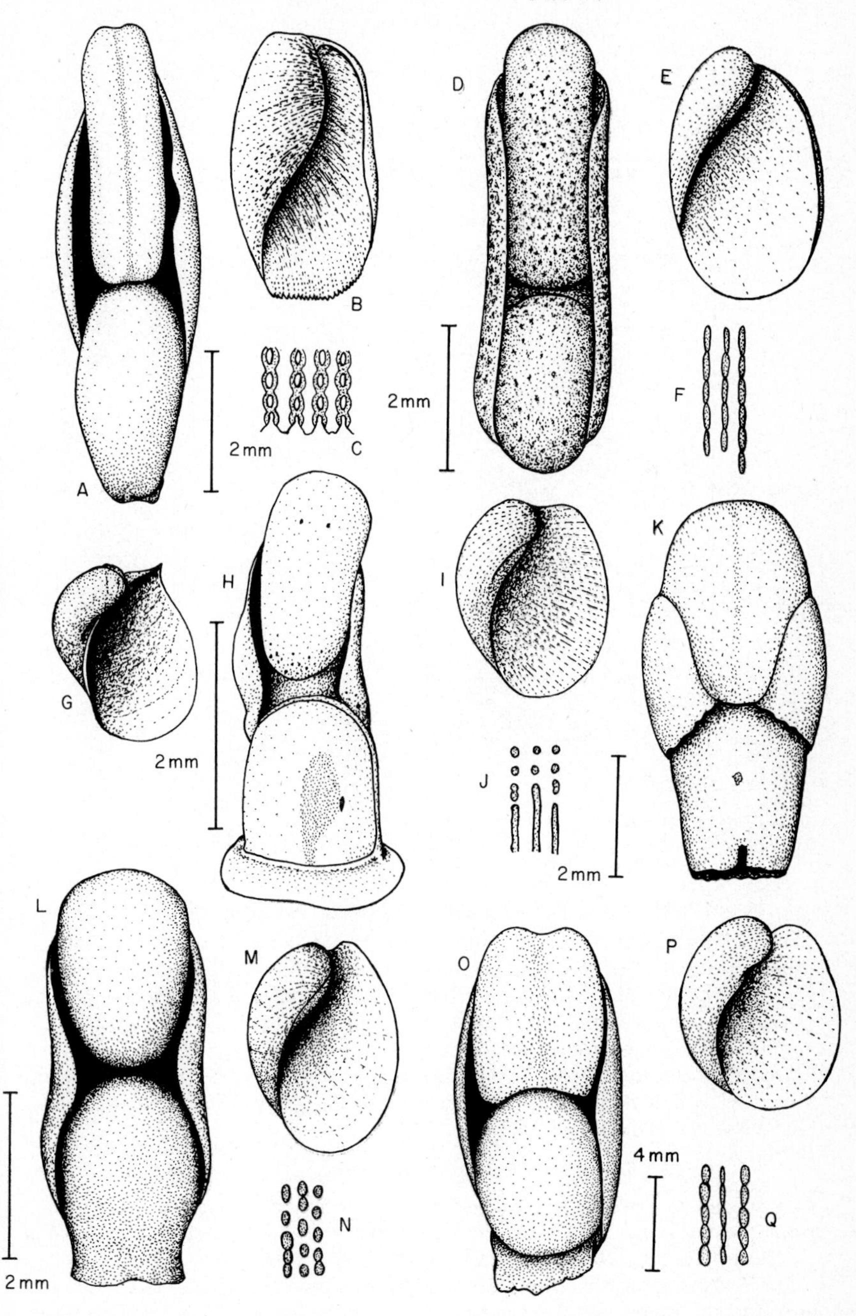
A
B
C
2mm
D
2mm
E
F
G
2mm
H
I
J
2mm
K
L
2mm
M
N
O
4mm
P
Q

Family PHILINOGLOSSIDAE

Genus PHILINOGLOSSA Hertling, 1932

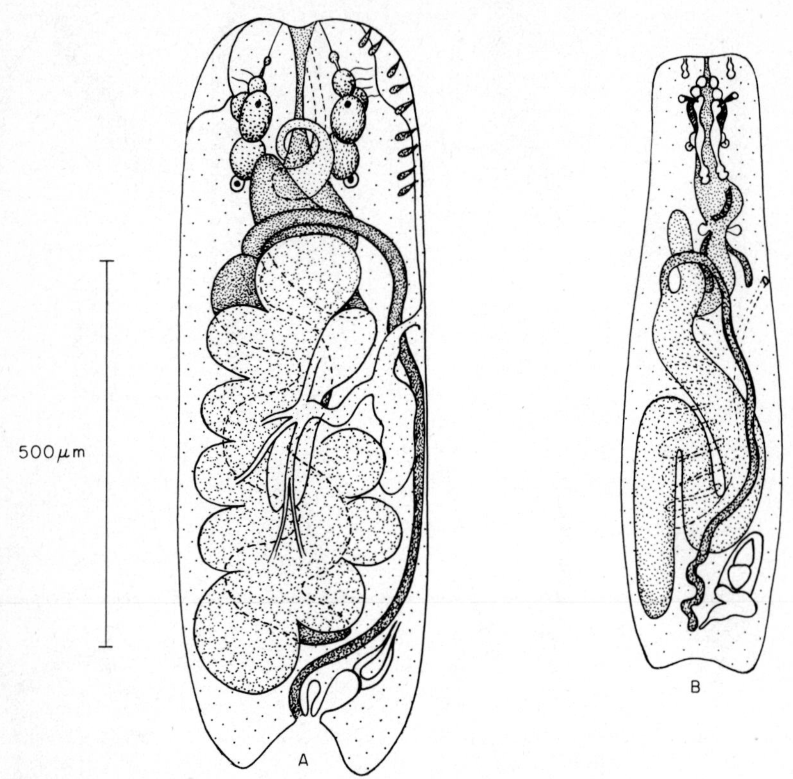

FIG. 14. *Philinoglossa remanei:* A, dorsal view; *Philinoglossa helgolandica:* B, dorsal view (after Hertling (1932) and Marcus & Marcus (1954)).

Philinoglossa helgolandica Hertling, 1932 (Fig. 14B)

This shell-less species measures up to 3 mm in extended length. The body is translucent grey in colour, with on the head a pair of darker longitudinal bands; the edges of the body are darker than the remainder. Black specks may be evident on the dorsum. Small glandular structures can be seen through the skin and are especially characteristic.

On British coasts this species is known at the present time only from Plymouth Sound where it is found in sub-littoral shelly gravel; further distribution Helgoland and Banyuls-sur-Mer.

Philinoglossa remanei Marcus and Marcus, 1958 (Fig. 14A)

This species resembles *P. helgolandica* fairly closely, but may be distinguished from the latter by the presence in *P. remanei* of well marked indentations anteriorly and posteriorly (compare Figs 14A and 14B).

On British shores this little-known species has been recorded only from shelly gravel in the Menai Straits; elsewhere it has been found in the Mediterranean Sea at Naples and Banyuls-sur-Mer.

Family RUNCINIDAE

Genus RUNCINA Forbes, 1853

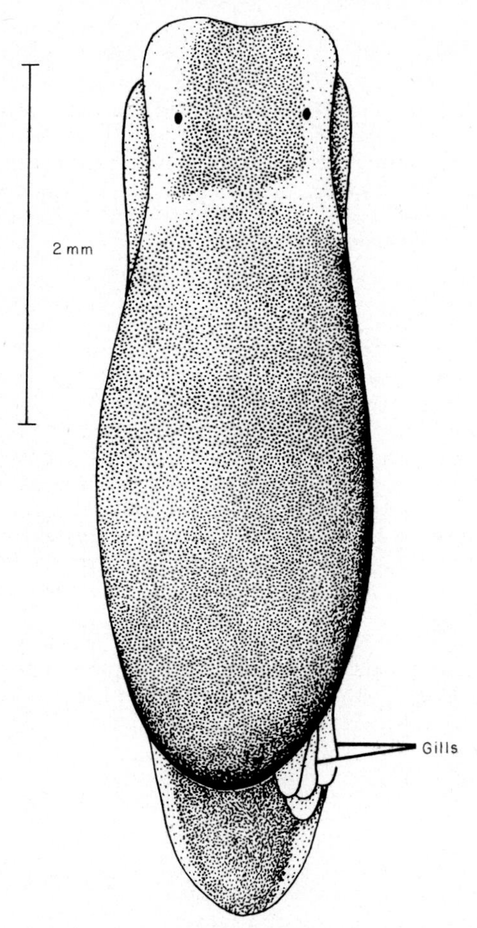

FIG. 15. *Runcina coronata* dorsal view.

Runcina coronata (Quatrefages, 1844) (Fig. 15)

Pelta coronata Quatrefages, 1844
Runcina hancocki Forbes and Hanley, 1851
R. calaritana Colosi, 1915

This is a shell-less species, up to 6 mm in overall length, dark brown, with paler areas on the head (Fig. 15). The animal is smooth and graceful and it has some resemblance to a sacoglossan sea-slug such as *Limapontia*, from which it can be distinguished, however, by the presence of gills under the rear lip of the mantle in *Runcina*.

This species occurs in clear coralline rock-pools all around the British Isles, feeding upon algae. Elsewhere it has been recorded from the Atlantic and Mediterranean coasts of France and from Portugal.

Family APLYSIIDAE

1. Parapodia joined high posteriorly; penis dark in colour **2**
   Parapodia separate posteriorly; penis white . . . *Aplysia fasciata* (p.40)
2. Pedal sole narrow; penis not black, spatulate without spines *Aplysia punctata* (p. 38)
   Pedal sole broad, often forming a circular metapodial sucker; penis black with spines visible when extended . . . . *Aplysia depilans* (p. 40)

Genus APLYSIA L., 1767

Aplysia punctata Cuvier, 1803 (Fig. 16A,D)

Aplysia rosea (Rathke, 1799)

Commonly known as the "sea hare". Shell internal (covered by the mantle), length up to 40 mm, transparent, fragile, pale amber in colour, the aperture wide, occupying nearly the whole of the ventral surface.

The body may reach 200 mm in overall length, olive green, brown or purplish black in colour (young specimens often red). The body-shape is characteristically long and narrow. The parapodial lobes are widely spaced in front, but joined rather high posteriorly. Swimming does not occur. If abruptly disturbed a slimy purple secretion can be emitted. Aplysiids are all herbivorous.

If doubt is experienced in identifying a difficult specimen, the penis must be dissected out and examined with a lens. This organ in *A. punctata* is broad, spatulate and lacking spines.

This is the most common British species of aplysiid and may be found in shallow water all around these islands, especially in the spring and early summer, when spawning occurs. Elsewhere it has been recorded from Greenland and Norway to the Mediterranean Sea and the Canary Islands.

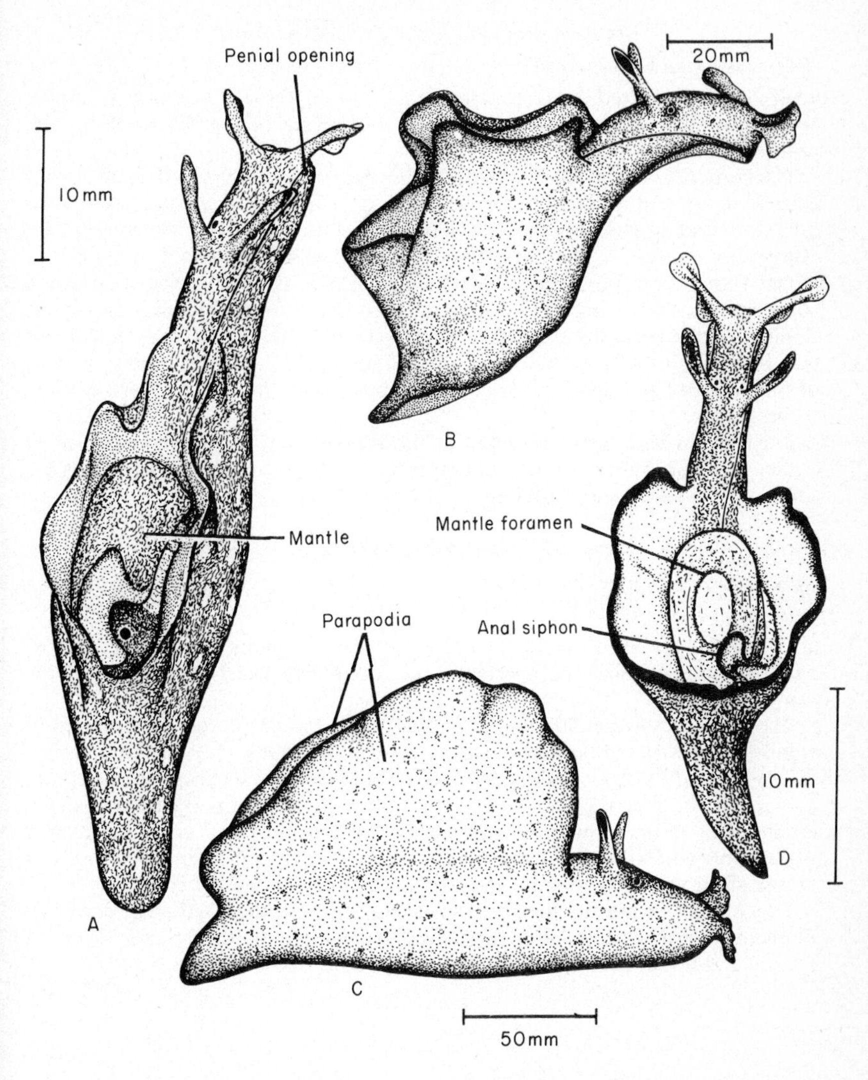

FIG. 16. *Aplysia punctata:* A, dorsal view; *Aplysia depilans:* B, lateral view; *Aplysia fasciata:* C, lateral view; *Aplysia punctata:* D, juvenile with black bordering.

Aplysia depilans Gmelin, 1791 (Fig. 16B)

Aplysia leporina Chiaje, 1822

Shell internal (covered by the mantle), length up to 40 mm, transparent, fragile, pale amber in colour, the aperture wide, occupying nearly the whole of the ventral surface.

The body may reach 300 mm in overall length, brown or greenish brown, with paler blotches and often with blackish veining. The body shape is characteristically low and broad and the rear part of the foot often dilates to form an attachment "sucker". The parapodial lobes are widely separated in front, but joined rather high posteriorly. Swimming has not been observed in Atlantic coast specimens, but certainly occurs in those found in the Mediterranean Sea. A slimy purple secretion is ejected from the mantle area if alarmed. *A. depilans* may be distinguished from the other aplysiids of northern Europe by examination of the penis. In this species it is stout and black with gross spiny warts around its basal part.

This species is only rarely found in the British Isles, on extreme south-west coasts. It is locally very common, however, on the coasts of France and as far south as West Africa and Madeira.

Aplysia fasciata Poiret, 1789 (Fig. 16C)

Aplysia limacina Blainville, 1823

A. depilans Blainville, 1823

Shell internal (covered by the mantle), length up to 70 mm, transparent, fragile, pale amber in colour, the aperture wide, occupying nearly the whole of the ventral surface.

The body may reach 400 mm in overall length, black or very dark brown in colour, often with red borders to the parapodia. The body-shape is characteristically high and narrow. The parapodial lobes are widely separated in front and behind; at the rear they unite very low down so that the mantle cavity gapes more widely than in the other two species. *A. fasciata* has a greatly elongated white penis which bears no spines. This species is a good swimmer. Floods of purple slime are released from the mantle cavity if the animal is molested.

This species is rare in Britain, and occurs only on the south-west coasts of England, Wales and Ireland. It is locally common on the atlantic coast of France and in the Mediterranean Sea as well as in West Africa.

Family PLEUROBRANCHIDAE

1. Mantle brown, papillate . . . . . *Pleurobranchus membranaceus* (p. 42)
Mantle yellow or orange, smooth **2**

2. Shell more than $\frac{1}{3}$ body-length . . . . . *Berthella plumula* (p. 44)
Shell less than $\frac{1}{4}$ body-length . . . . . . *Berthellina citrina* (p. 44)

Genus PLEUROBRANCHUS Cuvier, 1804

Pleurobranchus membranaceus (Montagu, 1815) (Fig. 17A-D)

Lamellaria membranacea Montagu, 1815

Pleurobranchus tuberculatus Meckel, 1808

Shell (Fig. 17D) internal (covered by the mantle), length up to 50 mm, transparent, fragile, the aperture wide, occupying nearly the whole of the ventral surface.

The body may reach 120 mm in overall length, pale brown, with patches of darker brown between the retractile, soft, conical mantle tubercles (Fig. 17A). The skin can secrete defensive sulphuric acid if attacked. An elongated gill is present under the right side of the mantle skirt. At the rear of the body, a conspicuous metapodial gland (Fig. 17A) develops at a pre-sexual stage, when the body is about 4 cm long. Anteriorly there are 2 pairs of head tentacles, the more dorsal of which (the rhinophores) are longitudinally enrolled (Fig. 17B). This species is a good swimmer, using asynchronous strokes of the sides of the dilated foot (Fig. 17C), the whole body fluttering into an upside-down position. It is specialized for feeding upon ascidians, into the outer skin of which it drills circular, straight-sided holes.

This is a common species at certain (unpredictable) times of year all around the British Isles; further distribution from the shores and shallow waters of France and Spain to the north African coast, down to 70 m depth.

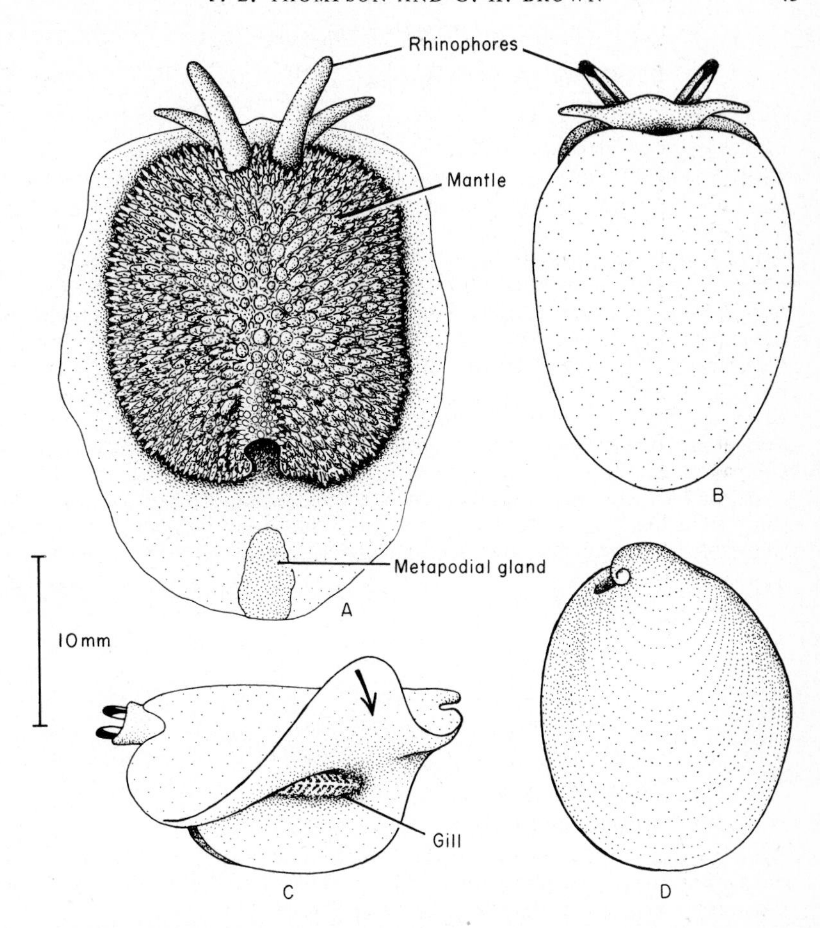

FIG. 17. *Pleurobranchus membranaceus:* A, dorsal view; B, ventral view; C, swimming individual; D, shell.

Genus BERTHELLA Blainville, 1824

Berthella plumula (Montagu, 1803) (Fig. 18A–D)

Bulla plumula Montagu, 1803

Berthella porosa Blainville, 1825

Shell (Fig. 18D) internal (covered by the mantle), length up to 30 mm, transparent, fragile, the aperture wide, occupying nearly the whole of the ventral surface.

The body may reach 60 mm in overall length, pale lemon-yellow to orange, often with reticulate markings in the middle of the smooth mantle. The skin can secrete defensive sulphuric acid if attacked. An elongated gill is present under the right side of the mantle skirt (Fig. 18C). The anal opening is situated in front of the rear of the gill mesentery, about half-way along the gill. At the rear of the body, a conspicuous metapodial gland (Fig. 18B) develops at a pre-sexual stage when the body is 175–30 mm long. Anteriorly there is a pair of enrolled rhinophoral tentacles above the flattened oral veil. It is specialized for feeding upon compound ascidians.

This is a common species, found under stones and in clear pools on the lower shore all around the British Isles. Elsewhere it has been recorded from Norway to the Mediterranean Sea, always in shallow water, down to 10 m.

Genus BERTHELLINA Gardiner, 1936

Berthellina citrina (Rüppell and Leuckart, 1828) (Fig. 18E and F)

Pleurobranchus citrina Rüppell and Leuckart, 1828

Berthellina engeli Gardiner, 1936

Shell internal (covered by the mantle), length up to 7 mm, transparent, fragile, the aperture wide, occupying nearly the whole of the ventral surface. It overlies the anterior part of the digestive gland (see Fig. 18F).

The body may reach 30 mm in extended length, yellow to orange in colour. The skin can secrete defensive sulphuric acid if attacked. An elongated gill is present under the right side of the mantle skirt. The anal opening is situated at the level of the rear of the gill mesentery near the tip of the gill (Fig. 18E). At the rear of the body, a conspicuous metapodial gland develops at a pre-sexual stage. Anteriorly there is a pair of enrolled rhinophoral tentacles above the flattened oral veil. It is not always easy to distinguish on external features alone between *Berthella plumula* and *Berthellina citrina*, but in the latter the internal shell is very much smaller in proportion to the body (Fig. 18). The diet is probably compound ascidians in both species.

This is a rare visitor to our southern shores, although it is prevalent on the Atlantic and Mediterranean coasts of France. Under various names it has been reported from various exotic localities, even as far away as Australia.

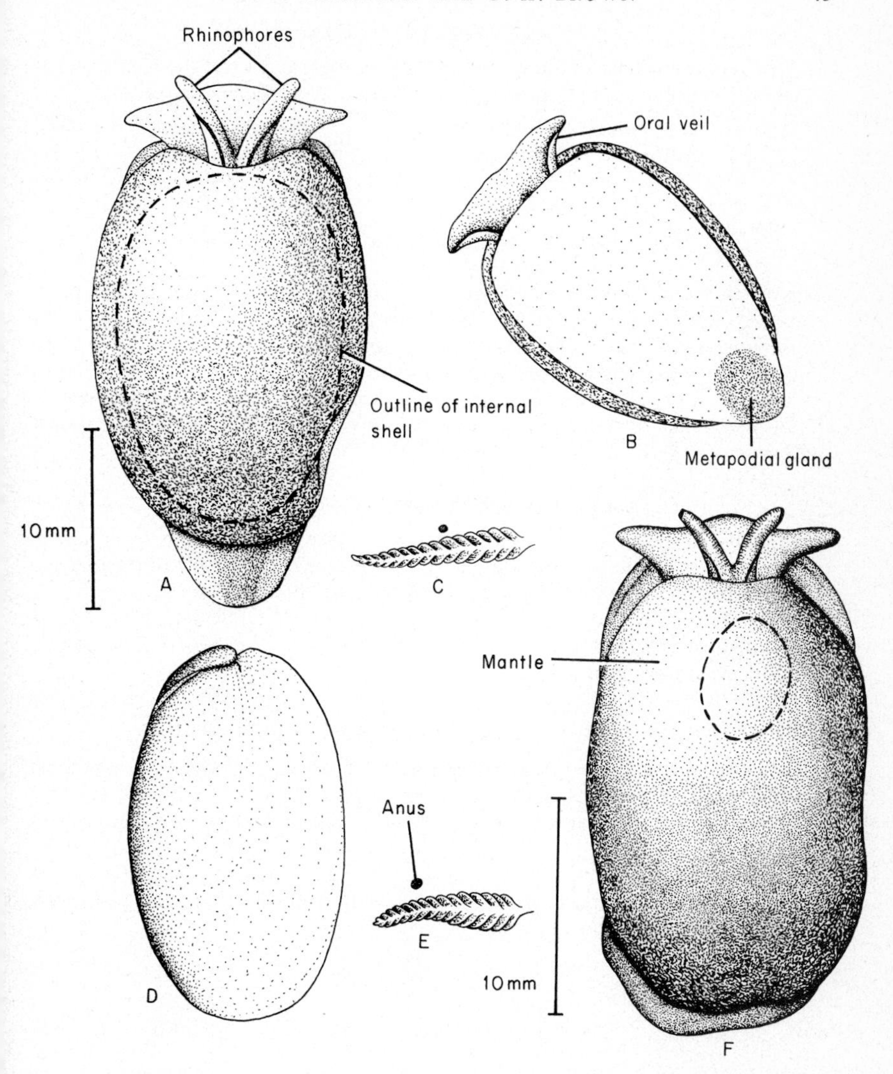

FIG. 18. *Berthella plumula:* A, dorsal view; B, ventral view; C, gill; D, shell; *Berthellina citrina:* E, gill; F, dorsal view.

Family HEDYLOPSIDAE

1. Needle-like spicules abundant in the mantle, present also on the head . 2
   Needle-like spicules present in the mantle only . *Hedylopsis suecica* (p. 46)
2. Anterior head tentacles cylindrical . . . . *Hedylopsis brambelli* (p. 46)
   Anterior head tentacles flattened . . . . *Hedylopsis spiculifera* (p. 46)

Genus HEDYLOPSIS Thiele, 1931

Hedylopsis brambelli Swedmark, 1968 (Fig. 19C)

This shell-less species measures up to 2·5 mm in extended length. The body colour is pale, often slightly pink or brownish. There are needle-like spicules densely arranged over the visceral mass, with a few also on the head. Eyes are said to be absent. The oral tentacles are cylindrical.

The only British records of this species are from North Wales where it is found in sublittoral shelly gravel; elsewhere it is known from the Atlantic coast of Sweden.

Hedylopsis spiculifera Kowalevsky, 1901 (Fig. 19B)

This shell-less species measures 1–2 mm in extended length. The body colour is brownish. Needle-like spicules are present in the mantle and in the oral tentacles. Eyes are present. The oral tentacles are broad and flattened.

The only British record of this species is from a locality off the Lizard, Cornwall, where it is found in fine shelly, sublittoral gravel; elsewhere it is known from the Mediterranean Sea.

Hedylopsis suecica Odhner, 1937 (Fig. 19A)

This shell-less species may reach 4 mm in extended length. The body colour is opaque white, sometimes with an amber tinge ventrally. Needle-like spicules are present only in the mantle. Eyes are present. The oral tentacles are broad and flattened.

The only British record of this species is from Plymouth Sound where it occurs in shelly sublittoral gravel; elsewhere it is known from the Skagerrak and from the Mediterranean Sea.

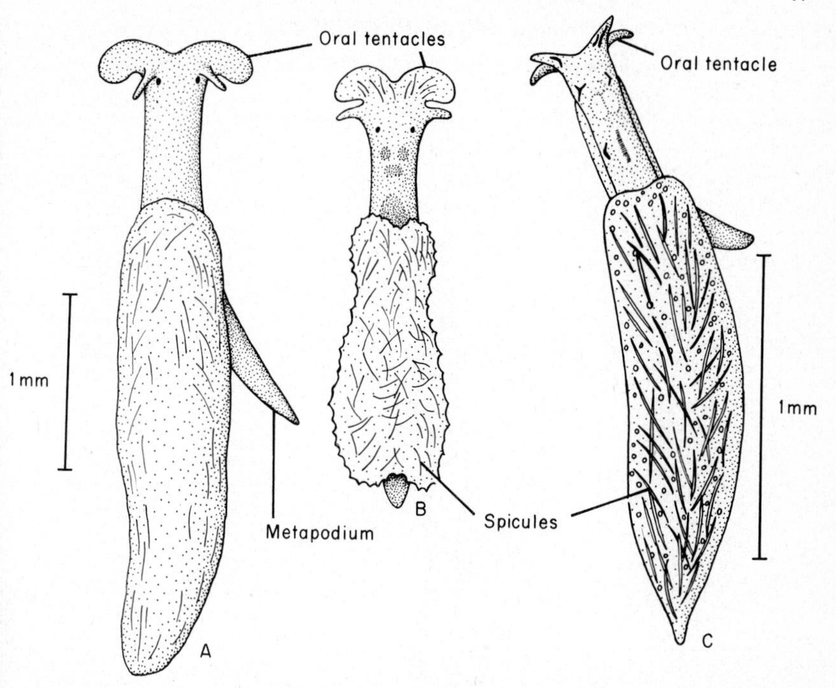

FIG. 19. *Hedylopsis suecica:* A, dorsal view; *Hedylopsis spiculifera:* B, dorsal view; *Hedylopsis brambelli:* C, dorsal view (after Odhner (1938), Kowalevsky (1901) and Swedmark (1968)).

Family MICROHEDYLIDAE

Genus MICROHEDYLE Hertling, 1930

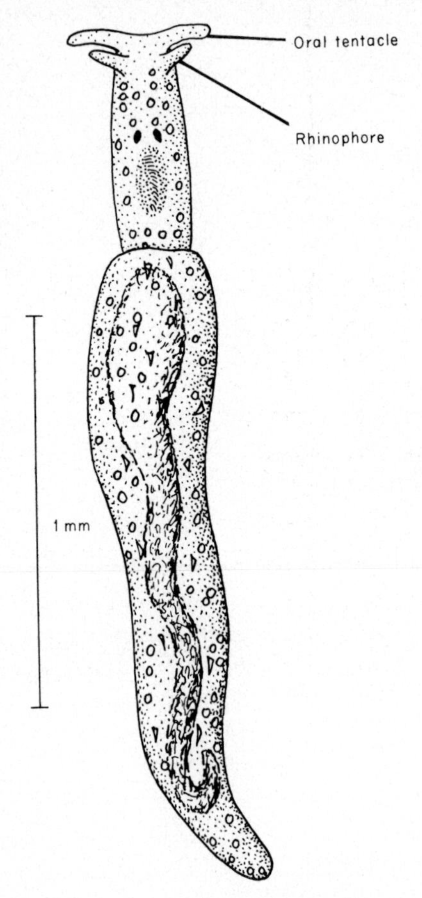

FIG. 20. *Microhedyle lactea* dorsal view (after Marcus and Marcus (1955)).

Microhedyle lactea Hertling, 1930 (Fig. 20)

This is another shell-less species and may reach 2 mm in extended length. The body colour is white. Stellate spicules are present in the mantle. Eyes are present. The oral tentacles are cylindrical.

This species, living in shelly gravel, has been recorded from Plymouth Sound and from a sublittoral locality off the Lizard, Cornwall. Further distribution: Helgoland, Mediterranean Sea.

Family STILIGERIDAE

1. Rhinophoral tentacles absent or forming short ear-shaped projections *Alderia modesta* (p. 52)
Rhinophores elongated and prominent **2**

2. Rhinophores enrolled **3**
Rhinophores cylindrical, not enrolled (but sometimes longitudinally grooved) . . . . *Stiliger bellulus* (p. 52)

3. Body-colour greenish . . . . *Hermaea dendritica* (p. 50)
Body-colour reddish **4**

4. Body with rose-red markings . . . . *Hermaea bifida* (p. 50)
Body with orange and purple markings . . . *Hermaea variopicta* (p. 50)

Genus HERMAEA Lovén, 1844

Hermaea bifida (Montagu, 1815) (Fig. 21A)

Doris bifida Montagu, 1815

The elongated body reaches up to 20 mm in length and is coloured white, with a rosy blush on the rhinophores and elsewhere on the head. Dorso-lateral cerata are arranged in up to 14 indistinct rows. Each ceras is elongate, pointed at the tip and faintly tuberculated. Brilliant red-brown digestive gland channels are visible through the translucent body wall and within the cerata. The metapodium is long and slender. The rhinophores are enrolled. The anal opening is antero-dorsal, just behind the head.

This delicate species is usually found on red algae such as *Griffithsia*, *Delesseria* or *Heterosiphonia*, rarely on the shore, more often in shallow water offshore, all around the British Isles. It has also been recorded from the Atlantic and Mediterranean coasts of France.

Hermaea dendritica (Alder and Hancock, 1843) (Fig 21B)

Calliopoea dendritica Alder and Hancock, 1843

The length of the body may reach 11 mm. It is coloured greenish white, with the digestive gland channels showing green or brownish through the translucent skin, especially within the cerata, which are arranged in a latero-dorsal pattern of up to 8 transverse rows. Each ceras is smooth and rounded at the tip. The anal opening is antero-dorsal, just behind the head. The rhinophores are enrolled.

This species lives and feeds upon the green algae *Codium* and *Bryopsis* and so most records are sublittoral, with rare captures on the lower shore, around all parts of the British Isles. Further distribution in N.W. Europe from Norway to the Mediterranean Sea, New England to the Caribbean Sea, California, Japan and eastern Australia.

Hermaea variopicta (Costa, 1869) (Fig. 21C and D)

Hermaeopsis variopicta Costa, 1869

This little known species has been recorded only once from Britain and that was from the lower shore among red algae at Thurlestone, Devon in 1972. Patches and streaks of orange and purple make this one of the most vividly patterned members of the Sacoglossa. As well as the pattern on the head shown in Fig. 21C, there is a system of concentric rings of orange and purple over the heart prominence in the middle of the dorsum. Each ceras (Fig. 21D) is purple in colour, with an orange tip, and white streaks fore and aft. The anal opening is antero-dorsal, just behind the head.

Hermaea variopicta has been recorded as rare from Brest and from the Mediterranean Sea and thus more information about this species will be most welcome.

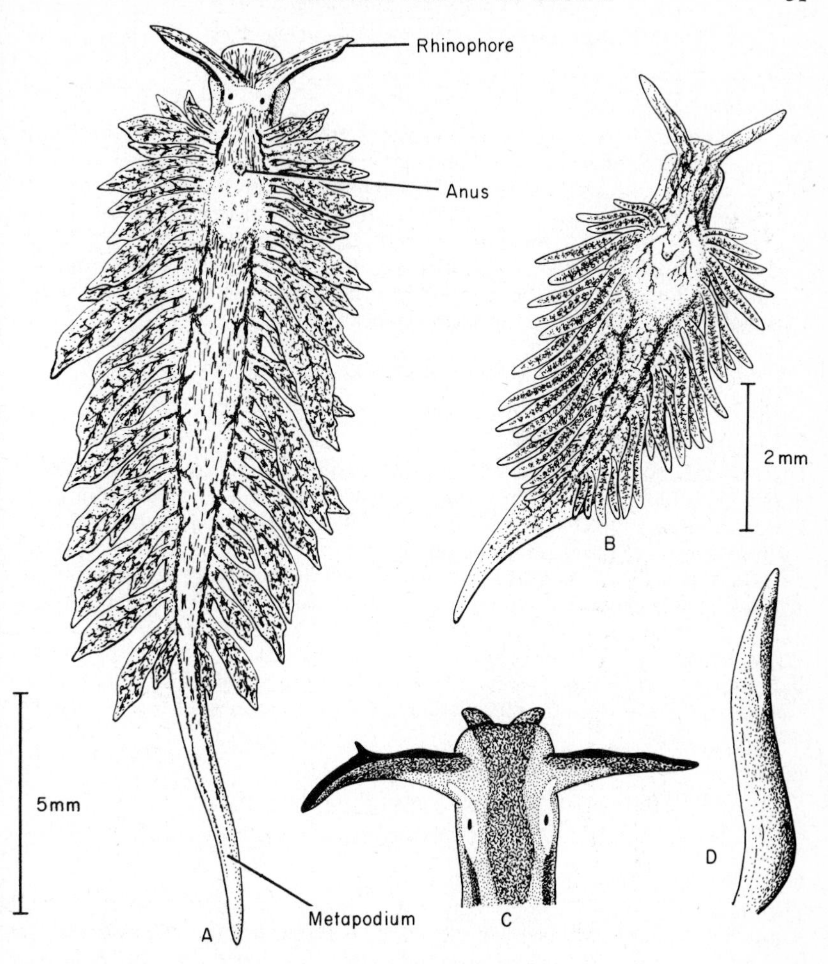

FIG. 21. *Hermaea bifida:* A, dorsal view; *Hermaea dendritica:* B, dorsal view (after Alder and Hancock (1845–55)); *Hermaea variopicta:* C, head, dorsal view; D, ceras.

Genus STILIGER Ehrenberg, 1831

Stiliger bellulus (Orbigny, 1837) (Fig. 22A)

Calliopoea bellula Orbigny, 1837

Embletonia mariae Meyer and Möbius, 1865

The body may reach a length of 10 mm and is greyish, marbled with white while the tail is pale red and the rhinophores grey. The cerata are arranged in 5 or 6 rows along the sides of the dorsum. The anal opening is antero-dorsal.

This uncommon species is found in shallow water, perhaps associated with *Zostera* among fronds of which it was first recorded in the Bassin d'Arcachon. It does not appear to live on saltmarshes like many other sacoglossans. In Britain and Ireland it has been reported only from sourthern coasts; further distribution from Norway to the Mediterranean Sea.

Genus ALDERIA Allman, 1846

Alderia modesta (Lovén, 1844) (Fig. 22B)

Stiliger modestus Lovén, 1844

The flattened body may reach 10 mm in extended length and is pale fawn in colour, with blotches of green, brown and white, the whole effect being drab. The cerata are arranged in up to 7 rows and regular muscular pulsations of these (visible with a lens) bring about blood-circulation; no heart is present. Through the skin of the cerata and of the foot can be seen the ramifying tributaries of the digestive gland. The anal opening is postero-dorsal (Fig. 22B).

This tough little sacoglossan has been recorded all around the British coasts, always associated with *Vaucheria* on saltmarshes. It is resistant to salinity changes within the range 5–36‰. It has a wide geographical distribution, from Norway to the Mediterranean Sea and along the Pacific coast of North America.

Family ELYSIIDAE

Genus ELYSIA Risso, 1818

Elysia viridis (Montagu, 1804)

Laplysia viridis Montagu, 1804 (Fig. 22C)

The flimsy, leaf-shaped body may reach 45 mm in length and the lateral parapodial "wings" may be outstretched or, as shown in Fig. 22C, held over the dorsum. The body-colour varies with the diet, from green to red. Relatively constant features, however, are tiny glistening red, blue and green spots; white patches may also be found, particularly at the edges of the parapodia, and black markings are sometimes seen on the head and elsewhere. Ramifying diverticula of the digestive gland and of the albumen gland are often visible through the skin of the parapodia and even of the rhinophores and other parts of the head. The anal opening is antero-lateral, below the right rhinophore. The rhinophores are enrolled.

This common species occurs on a variety of shallow-water algae, especially *Codium* and *Cladophora*, all around the British Isles. Elsewhere it has been recorded from Norway to the Mediterranean Sea, and there is a doubtful report from Chinese waters.

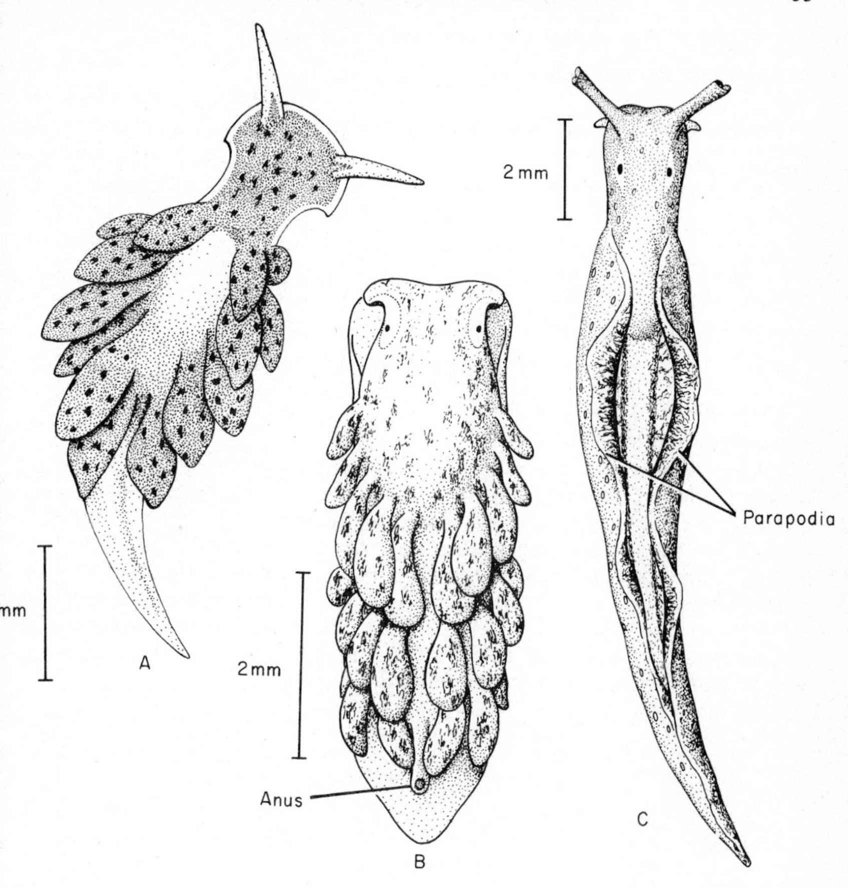

FIG. 22. *Stiliger bellulus* A, dorsal view (after Meyer and Möbius (1865)); *Alderia modesta:* B, dorsal view; *Elysia viridis:* C, dorsal view.

Family LIMAPONTIIDAE

1. Saltmarsh dwellers; head with no trace of tentacles *Limapontia depressa* (p. 56)
Coralline rockpool dwellers; head with ear-like or finger-like tentacles . **2**

2. Head tentacles ear-like . . . . *Limapontia capitata* (p. 54)
Head tentacles finger-like . . . . *Limapontia senestra* (p. 56)

Genus LIMAPONTIA Johnston, 1836

Limapontia capitata (Müller, 1774) (Fig. 23D)

Fasciola capitata Müller, 1774

Limapontia nigra Johnston, 1836

The smooth body may reach 8 mm in extended length, but usually does not exceed 4 mm, dark brown or black in colour, with paler areas around the eyes. Ear-shaped rhinophoral crests or ridges are present on the sides of the head (Fig. 23D). The metapodium is elongated, slender and usually pale. The anal and renal openings are situated near the mid-line, very close together.

This little sacoglossan may be found in coralline pools around the middle of the shore in clean coastal areas all round the British Isles. It feeds upon species of *Cladophora*, *Enteromorpha* and *Bryopsis*. Further distribution from the White Sea, Iceland, Norway to the Mediterranean Sea.

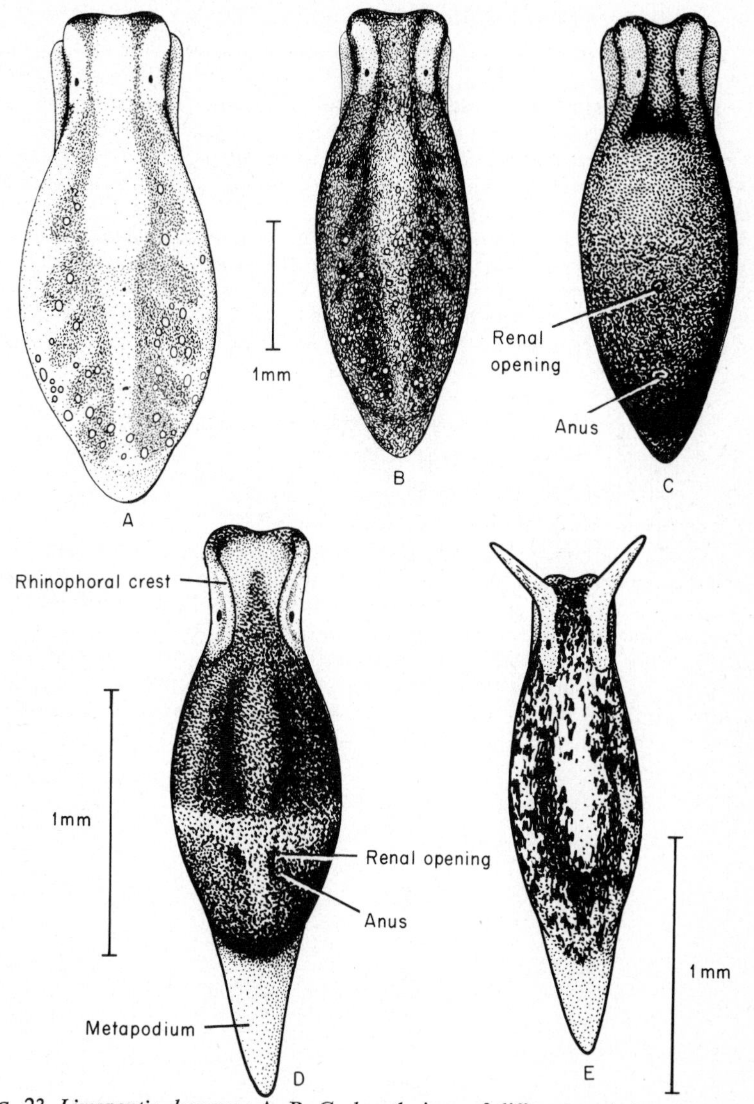

FIG. 23. *Limapontia depressa:* A, B, C, dorsal views of different varieties; *Limapontia capitata:* D, dorsal view; *Limapontia senestra:* E, dorsal view.

Limapontia depressa Alder and Hancock, 1862 (Fig. 23 A–C)

The smooth body may reach 6 mm in extended length, in colour usually dark brown or black (Fig. 23B and C), but in the variety *pellucida* Kevan the body is yellow with the greenish digestive gland visible through the translucent skin (Fig. 23A). The head bears no trace of tentacles, distinguishing this species at a glance from *L. capitata* and *L. senestra*. Additionally, the body-shape of *L. depressa* is more flattened and the dorsally situated anal and renal openings are some considerable distance apart (Fig. 23C).

This species is a characteristic inhabitant, with *Alderia modesta*, of British saltmarshes, as well as those of the continent of Europe; it is rarely found in pools, but usually on damp mud adjacent to the moister places, crawling and feeding upon *Vaucheria*. The salinity tolerance is not so great as that of *Alderia* and specimens survive best in approximately 25‰ sea water. Further distribution from the Baltic Sea to the Mediterranean.

Limapontia senestra (Quatrefages, 1844) (Fig. 23E)

Acteonia senestra Quatrefages, 1844

Cenia cocksi Alder and Hancock, 1855

Cenia corrugata Alder and Hancock, 1855

The slug-like body may reach 6 mm, marbled olive-brown to black with paler extremities (metapodium and tentacles). In the variety *corrugata* Alder and Hancock, the dorsum is slightly wrinkled. On the head of the adult is a pair of finger-like rhinophoral tentacles, absent in other British species of *Limapontia*; these tentacles may be lacking, however, in juveniles shorter than 2 mm.

This species is found, often with *Limapontia capitata*, in fully saline coralline intertidal pools, all around the British Isles. Elsewhere, it has been recorded from Norway to the French Atlantic coast.

Family TRITONIIDAE

1. Body length up to 20 cm; oral veil bilobed, each lobe bearing numerous short digitiform processes . . . . *Tritonia hombergi* (p. 58)
Body length up to 34 mm; oral veil not markedly bilobed, but bearing a few (maximally 6) long digitiform processes **2**

2. Body white with opaque white dorsal longitudinal markings
Tritonia lineata (p. 62)
Body yellow, brown or pink, without such white markings **3**

3. Body yellow or brown . . . . *Tritonia plebeia* (p. 60)
Body bright pink . . . . *Tritonia odhneri* (p. 62)

Genus TRITONIA Cuvier, 1798

Tritonia hombergi Cuvier, 1803 (Fig. 24A and B)

Tritonia alba Alder and Hancock, 1854

The body length may reach 200 mm; this is the largest British nudibranch mollusc and the only British opisthobranch reported reliably to be harmful to man. Byne in the last century found its secretions to blister his hands; it would be interesting to collect information on this because the authors have been unable to find any corroboration. The body varies in colour from white to dark purplish brown (generally darker with age), lighter ventrally. The mantle is covered by soft tubercles; dorso-laterally it bears conspicuous pallial gills. Certain of these gills are larger than the others and are flexed towards the median plane of the body; smaller gills project out laterally. The number of individual gills increases markedly with age. In juveniles (Fig. 24B), formerly regarded as a separate species, *T. alba*, white pigment forms patches mesial to each pallial gill. The frontal margin of the head (the oral veil) is strongly bilobed even in young ones and each lobe is divided into numerous finger-like processes.

This tritoniid is always found in association with the soft coral *Alcyonium digitatum*, and occurs all around the British Isles, down to 80 m. The chitinous jaws of this nudibranch have been discovered in dogfish stomachs. Outside Britain it has been reported from Norway to the Mediterranean Sea.

FIG. 24. *Tritonia hombergi* dorsal views: A, veniladult; B, jue.

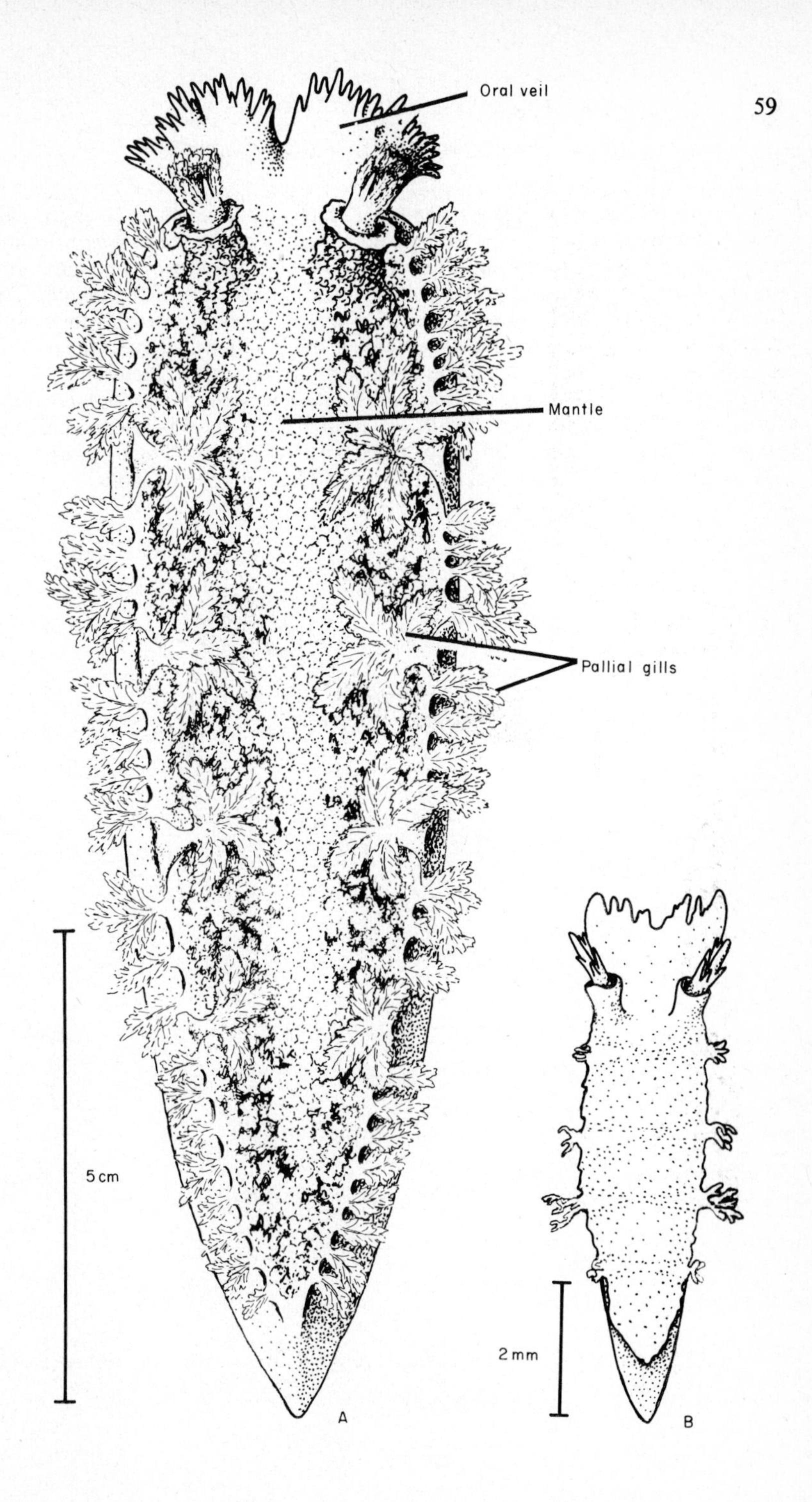
Oral veil
Mantle
Pallial gills
5 cm
2 mm
A
B

Tritonia plebeia Johnston, 1828 (Fig. 25A and B.)

The length of the body reaches 30 mm (usually not more than 20 mm), pale yellow to brownish (according to the colour-variety of the *Alcyonium* prey and subject to the generalization that the colour darkens with age). The pallial gills (up to 6 pairs) arise individually from the dorso-lateral edge of the mantle. The frontal margin of the head is divided into about 6 (rarely up to 8) finger-like projections. The general aspect of the body is more flattened than in the 3 other British species of *Tritonia*. Juveniles (Fig. 25B) are difficult to identify.

This tritoniid feeds upon the soft corals *Alcyonium digitatum* and *Eunicella verrucosa* and occurs in clear offshore water all around the British Isles, down to 60 m. Elsewhere it has been reported from Norway to Portugal and the Mediterranean Sea. The deepest record is from 150 m near the Faroes.

FIG. 25. *Tritonia plebeia* dorsal views: A, adult; B, juvenile (after Miller (1958)).

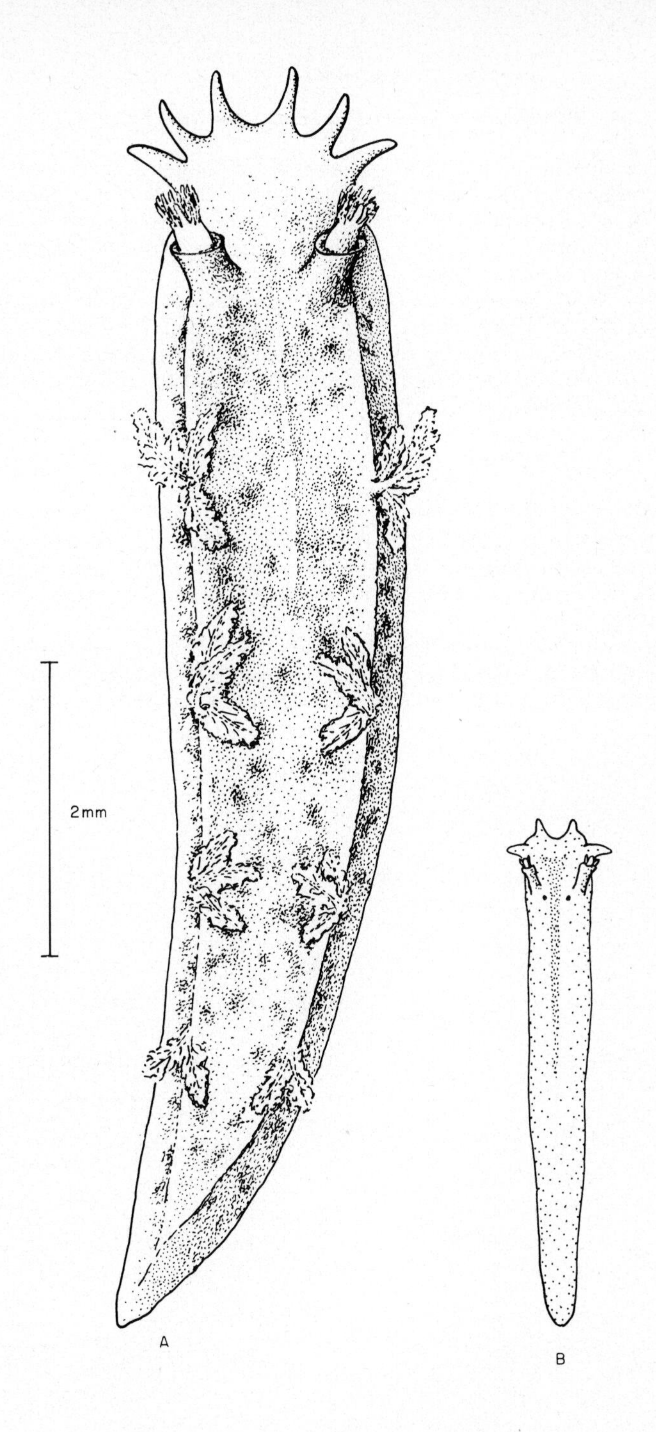
2mm
A
B

Tritonia lineata Alder and Hancock, 1848 (Fig. 26)

This species may attain a length of 34 mm, white (sometimes pinkish) in colour with a conspicuous line of opaque white down each side of the dorsum. The pallial gills (4–6 pairs) arise individually from the dorso-lateral edge of the mantle. The frontal margin of the head is divided into 4 elongated finger-like processes.

There is uncertainty about the diet of *T. lineata*. It has long been assumed that it fed upon *Alcyonium digitatum* (perhaps largely because it so closely resembles this prey), but aqualung divers have only rarely found *T. lineata* on or near *Alcyonium*. More information on its feeding habits is clearly needed. It is common in clean shallow offshore localities all around the British Isles; further distribution from Norway to Brittany down to 40 m.

Tritonia odhneri (Tardy, 1963) (Fig. 27)

Duvaucelia odhneri Tardy, 1963

The body length may reach 34 mm, rose pink in colour, like that of its prey, the gorgonian *Eunicella verrucosa*. Areas mesial to the pallial gills (up to 8 pairs) are paler. The frontal margin of the head is produced to form 6 elongated finger-like processes.

This interesting species has been recorded around Britain from shallow water off western Ireland, Jersey, Lundy and N. Cornwall. Elsewhere it is known only from the type locality near the Ile de Ré on the Biscay coast of France.

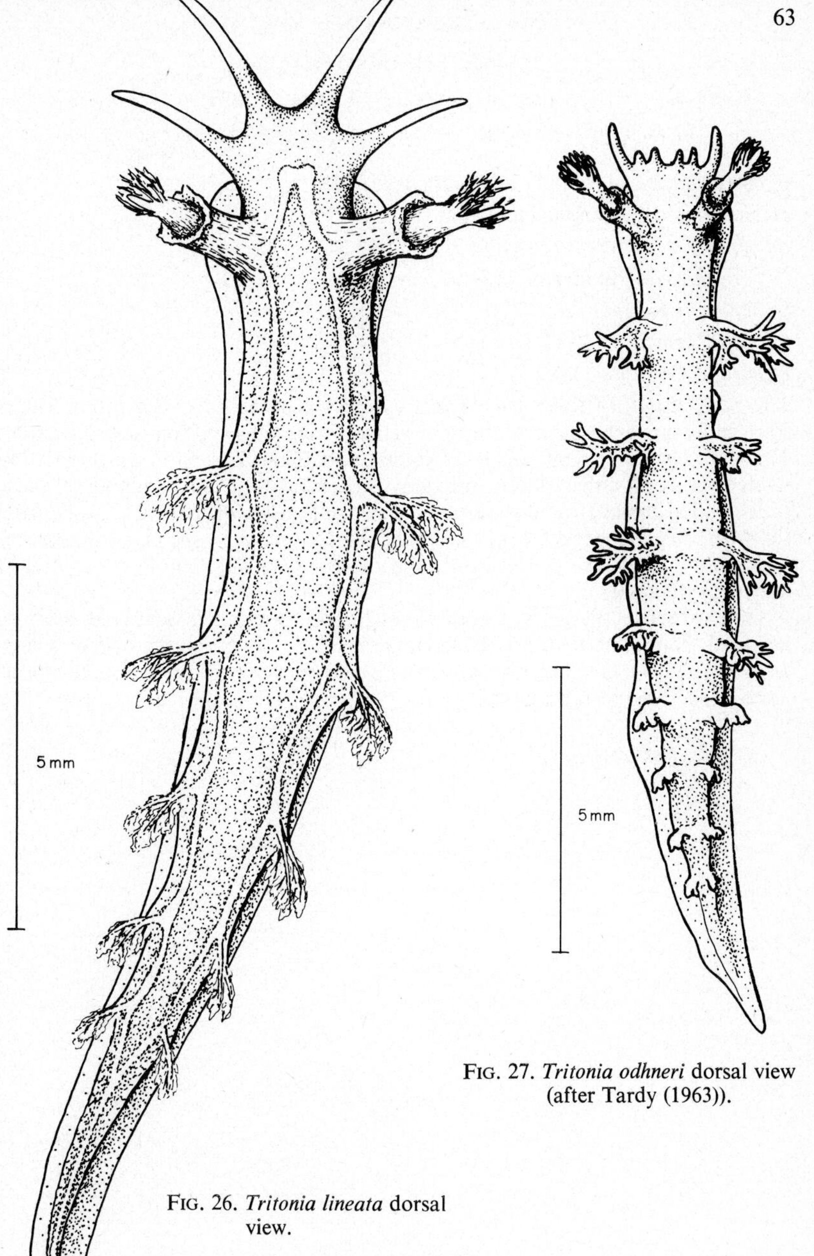

FIG. 27. *Tritonia odhneri* dorsal view (after Tardy (1963)).

FIG. 26. *Tritonia lineata* dorsal view.

Family LOMANOTIDAE

Genus LOMANOTUS Vérany, 1844

Lomanotus marmoratus (Alder and Hancock, 1845) (Fig. 28)

Eumenis marmoratas Alder and Hancock, 1845
E. flavida Alder and Hancock, 1846
Lomanotus genei Vérany, 1846
L. portlandicus Thompson, 1859
L. hancocki Norman, 1877
L. eisigii Trinchese, 1883
L. varians Garstang, 1889

The body length of this variable species may reach 55 mm. The colour varies from white through yellow to crimson; generally the paler specimens are juveniles. The dorso-lateral pallial rim is expanded and divided to form an undulating series of gills on either side of the body. Posteriorly the two longitudinal parts of the pallial rim are raised to form a tail which is lashed from side to side during the swimming escape reaction. The lamellate rhinophores issue from tall sheaths which have finger-like processes around the rim. The oral veil bears 2 pairs of finger-like tentacles.

This attractive species has been found to be common in scattered offshore areas all around the British Isles, feeding on sturdy calyptoblastic hydroids such as *Nemertesia antennina*, down to 40 m; it is also known from Brittany and various parts of the Mediterranean Sea.

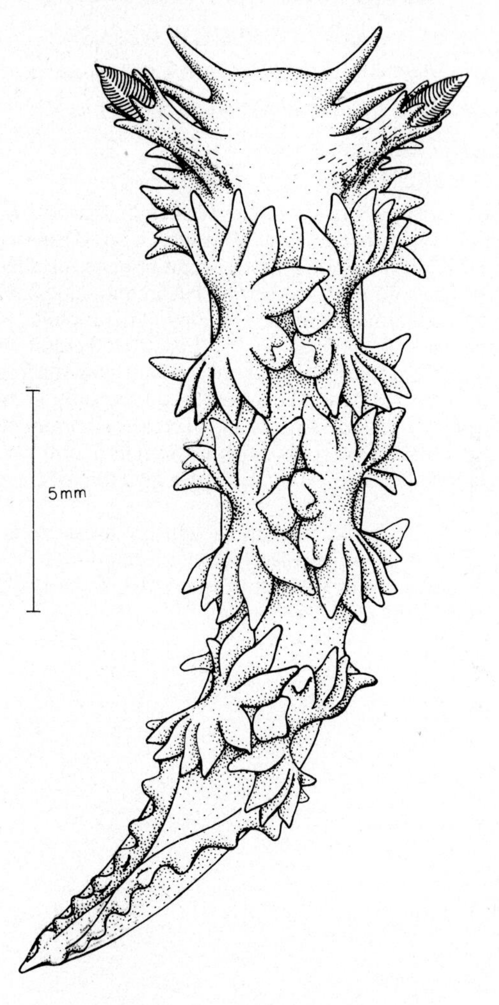

FIG. 28. *Lomanotus marmoratus* dorsal view.

Family DENDRONOTIDAE

Genus DENDRONOTUS (Alder and Hancock, 1845)

Dendronotus frondosus (Ascanius, 1774) (Fig. 29A and B)

Amphitrite frondosa Ascanius, 1774

Doris arborescens Müller, 1776

This handsome species may reach 100 mm in length, marbled with white, orange, red and brown in varying proportions. Juveniles up to a length of 4 mm are pale, then from 4–30 mm the developing patterns can be vivid forming body stripes and contrasting bands of colour on the cerata, but in individuals longer than 30 mm the pattern once again appears drab. The dorso-lateral pallial rim bears on each side of the body a series of up to 9 arborescent ceratal processes. Similar processes protrude from the rhinophore sheaths and from the frontal margin of the head. The foot is narrow and the whole body laterally compressed, and the body can be lashed from side to side in a feeble swimming escape reaction. It feeds upon calyptoblastic hydroids (such as *Sertularia* and *Hydrallmania*) when young, but transfers its attention to the larger gymnoblasts, especially *Tubularia indivisa*, when adult.

Dendronotus may be found in shallow offshore areas all around the British Isles. Elsewhere it has been recorded from Greenland and the Arctic Sea down to the French Atlantic coast, from both seaboards of the U.S.A., and there is a single record from China.

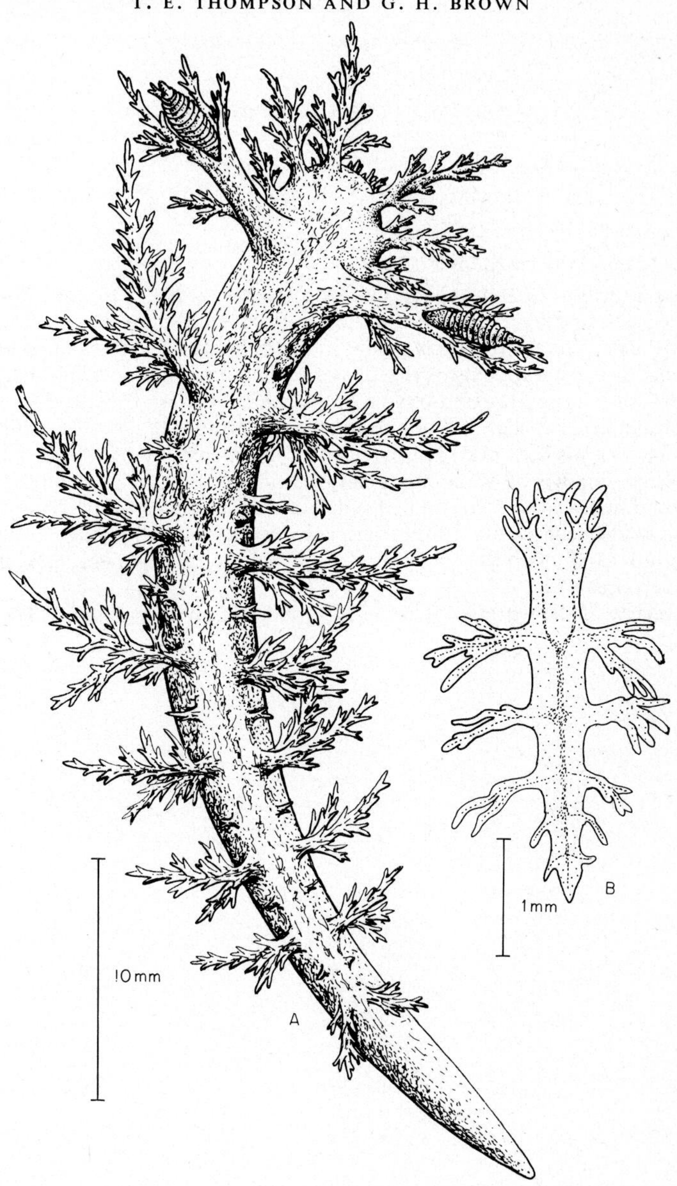

FIG. 29. *Dendronotus frondosus* dorsal views: A, adult; B, juvenile (after Miller (1958)).

Family HANCOCKIIDAE

Genus HANCOCKIA (Goose, 1877)

Hancockia uncinata (Hesse, 1872) (Fig. 30)

Doto uncinata Hesse, 1872

Hancockia eudactylota Gosse, 1877

Govia rubra Trinchese, 1885

Govia viridis Trinchese, 1885

The body length of this apparently rare species may reach 14 mm. The colour is pale green or pink, dotted with white and the overall shape is extremely slender and delicate. The dorso-lateral pallial rim is expanded at intervals, on each side forming 4–5 ceratal outgrowths of a peculiar and unique kind. Each ceras consists of a structure resembling a half-clenched human fist (see Fig. 30). The longitudinally lamellate rhinophores emerge from tall sheaths which have crenulate rims. The oral veil is bilobed, each lobe bearing 3–4 finger-like processes. Bundles of nematocysts can be discharged by the body (especially the cerata) if attacked. It feeds on hydroids but details are not available.

On British coasts this species has been found only in shallow water near Plymouth and Torbay; elsewhere it is known from Arcachon and the Mediterranean Sea.

A review of the world's species of this family has been published (Thompson, 1972).

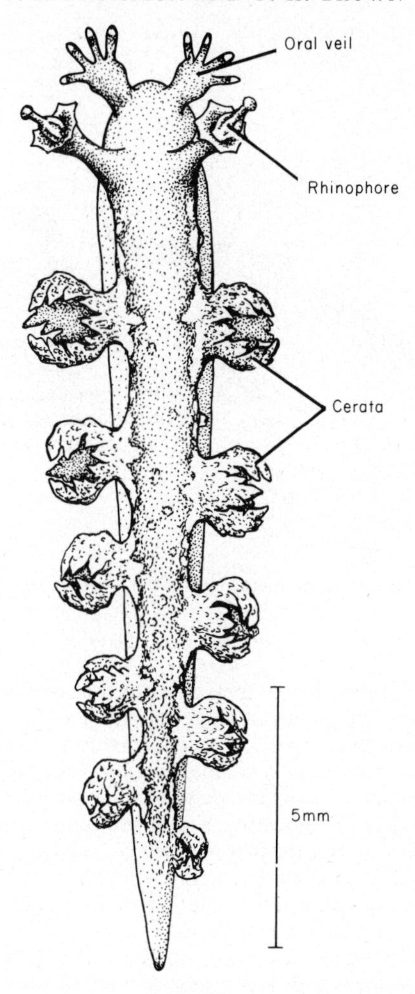

FIG. 30. *Hancockia uncinata* dorsal view (after Eliot (1912)).

Family SCYLLAEIDAE
Genus SCYLLAEA L., 1758

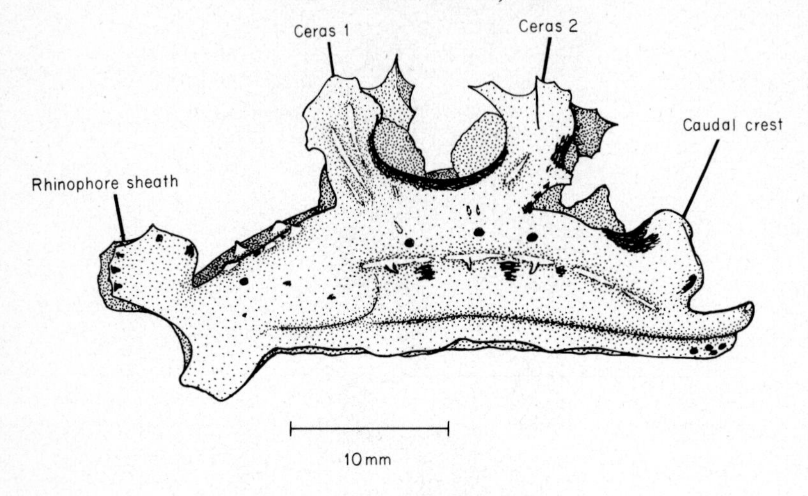

FIG. 31. *Scyllaea pelagica* lateral view (after Marcus and Marcus (1963)).

Scyllaea pelagica L., 1758 (Fig. 31)

The length of the body reaches 60 mm, pale green or brown in colour with green or blue lateral spots and scattered white markings which sometimes aggregate to form white streaks on the flanks. The whole body is laterally flattened and the foot is very narrow; a swimming escape reaction, with side to side lashing, can be elicited by a simulated attack. In side view, there are, on each side of the body, four large structures which resemble cerata, projecting conspicuously from the dorso-lateral pallial rim. But the first of these is a dilated rhinophore sheath and the last is a metapodial crest-like structure; the other two pairs are true cerata. All these processes are characteristically jagged in outline and may bear small gill-like excrescences on their mesial faces.

This species is very rare on British coasts although reputedly common in tropical seas around the world. It is always associated with floating brown algae (*Sargassum*, for instance), preying on attached hydroids. In this situation it is well camouflaged. The only British record relates to 3 individuals cast up on *Saccorhiza bulbosa* at Falmouth.

Family DOTOIDAE

1. Cerata evenly coloured, brownish 2
Cerata bearing tubercles which are tipped with colour (red, black or white) 3

2. Rhinophore sheaths smoothly flared . . . . . . *Doto fragilis* (p. 78)
Rhinophore sheaths crenelated . . . . *Doto cuspidata* (p. 73)

3. Ceratal tubercles tipped crimson . . . . *Doto coronata* (p. 74)
Ceratal tubercles tipped dark brown or black . . *Doto pinnatifida* (p. 76)
Ceratal tubercles tipped white . . . . *Doto cinerea* (p. 72)

Genus DOTO Oken, 1815

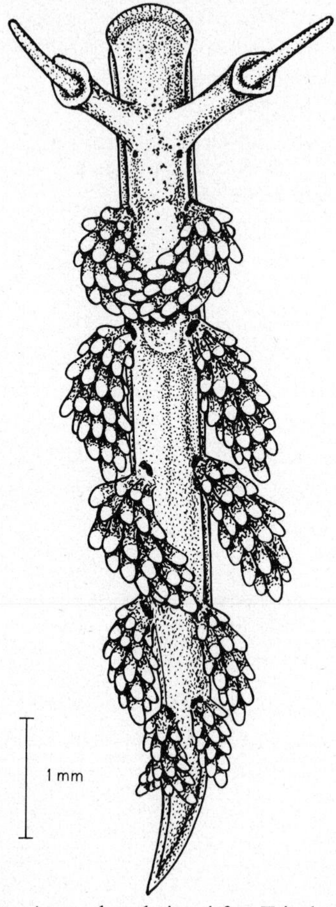

FIG. 32. *Doto cinerea* dorsal view (after Trinchese (1881)).

Doto cinerea Trinchese, 1881 (Fig. 32)

The body length may reach 8 mm, yellowish with blackish mottling above. The dorsal cerata (4–6 pairs) bear concentric rings of tubercles, each of which is brown but pale at the summit. The frontal margin of the head is smoothly rounded, not dilated or crested.

This is the least well known British species of *Doto* and more information is urgently needed about several features of the morphology. *D. cinerea* has been recorded only once (on *Sertularia*) from the British Isles, from Plymouth. Elsewhere it is known from the Atlantic and Mediterranean coasts of France, and from the Cape Verde Islands.

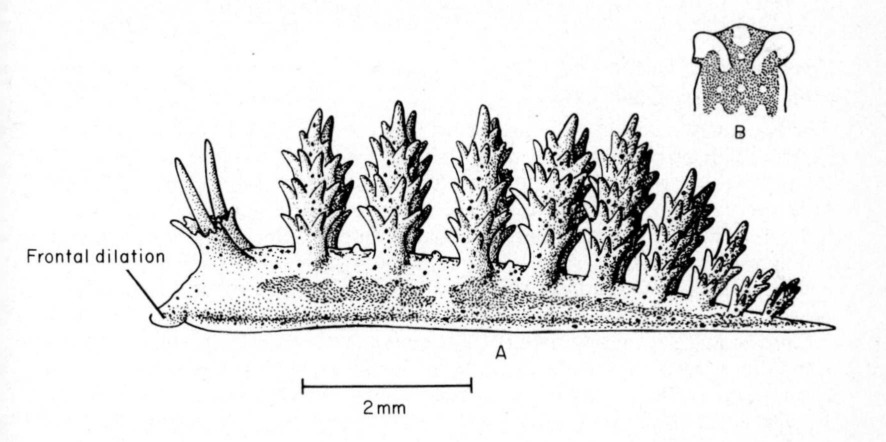

FIG. 33. *Doto cuspidata:* A, lateral view; B, head, dorsal view (after Miller (1958)).

Doto cuspidata Alder and Hancock, 1862 (Fig. 33A and B)

Doto cornaliae Trinchese, 1881

D. aurea Trinchese, 1881

The length of the body reaches 11 mm, pale fawn in colour with a dense speckling of dark brown. The dorsal cerata (up to 10 pairs) bear concentric rows of elongated tubercles which bear no especially conspicuous pigment markings. The dilated rhinophore sheaths are (at least in adults) crenulated. The frontal margin of the head is dilated laterally (Fig. 33B).

Alder and Hancock (1845–55) considered this species to exhibit many features intermediate between *D. coronata* and *D. fragilis*. It is nevertheless a clearly distinct and separate species.

In British waters it has been recorded only from Shetland and from the Isle of Man (on *Nemertesia ramosa*). Elsewhere it is known from the Mediterranean Sea and from Trondhjem Fjord, down to 160 m.

Doto coronata (Gmelin, 1791) (Fig. 34)

Doris coronata Gmelin, 1791

Doto costae Trinchese, 1881

The body may reach 15 mm in extended length (usually 10–12 mm), pale yellow or white with red or purplish pigment above, the spots of this pigment occasionally lining up to form streaks or blotches. The dorsal cerata (up to 8 pairs) are pale and bear concentric rings of tubercles, each of which is surmounted by a dark red spot. The dilated rhinophore sheaths are smooth edged. The frontal margin of the head is dilated laterally.

This species is common in shallow offshore waters, living and feeding on a variety of calyptoblastic hydroids (especially *Sertularia argentea*, *Obelia dichotoma* and *Plumularia setosa*), and sometimes also on gymnoblasts (such as *Clava multicornis*, *Tubularia larynx* and *Sarsia eximia*). It has been reported from localities all around the British Isles; further distribution from Murmansk, Iceland and Norway to Portugal and the Mediterranean Sea, and on the coast of N. America in the state of Maine, down to 200 m.

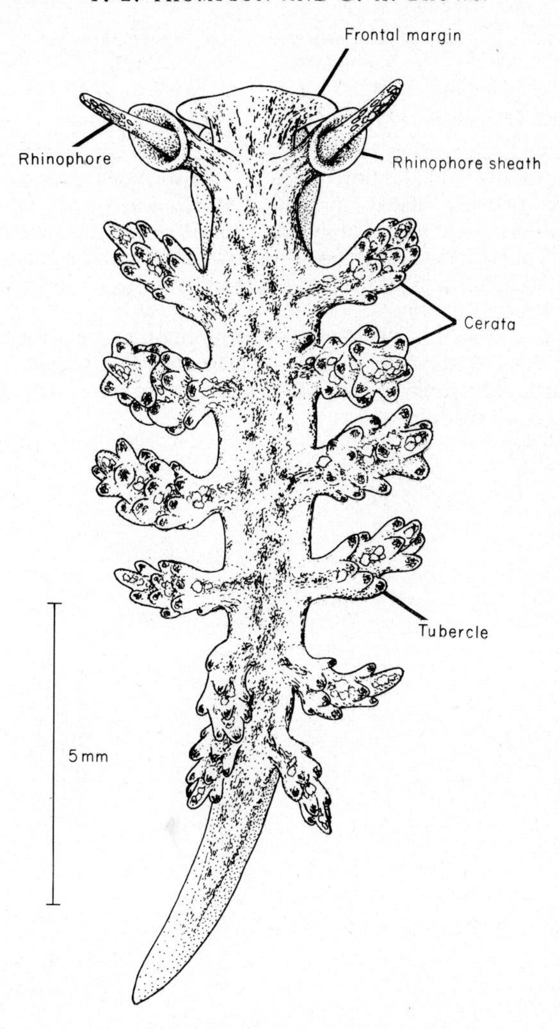

FIG. 34. *Doto coronata* dorsal view.

Doto pinnatifida (Montagu, 1804) (Fig. 35)

Doris pinnatifida Montagu, 1804

?*Doto splendida* Trinchese, 1881

The length of the body reaches 29 mm, pale fawn in colour with dark brown to black mottling above; in addition there are conspicuous dark brown spots around the rim of each dilated rhinophore sheath and on the tips of small tubercles down each side of the body. The dorsal cerata (up to 9 pairs) bear concentric rows of elongated tubercles each of which terminates in a dark brown or black spot. The coloration of the body darkens with age. The frontal margin of the head is dilated laterally.

This species is common in shallow water offshore on the southwestern coasts of the British Isles, feeding upon calyptoblastic hydroids (chiefly *Nemertesia antennina*). Further distribution New Hampshire on the north American coast and the Biscay coast of France.

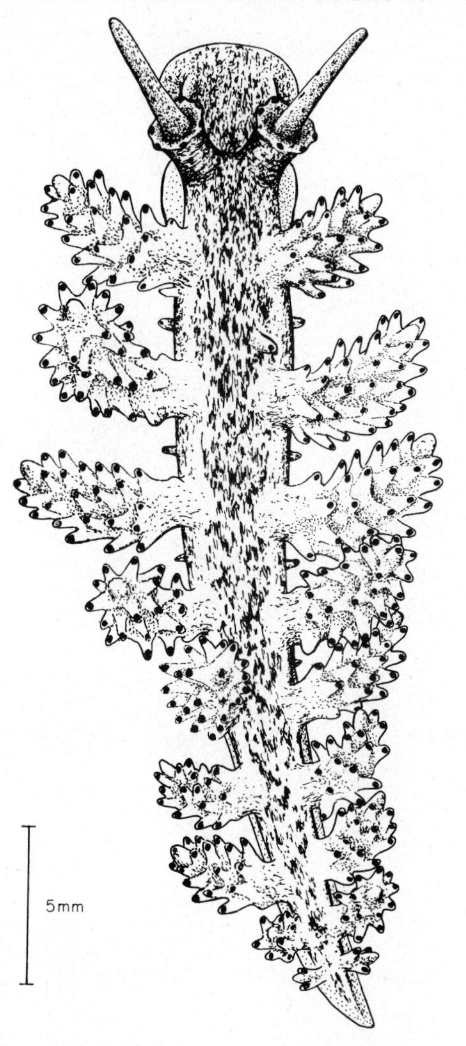

FIG. 35. *Doto pinnatifida* dorsal view

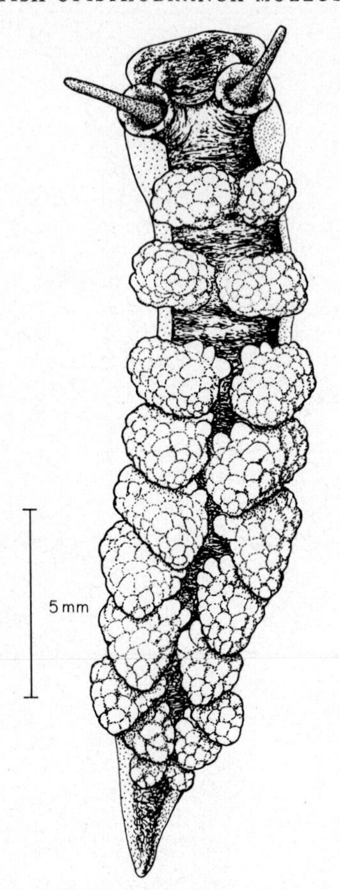

FIG. 36. *Doto fragilis* dorsal view.

Doto fragilis (Forbes, 1838) (Fig. 36)

Melibaea fragilis Forbes, 1838

The body length may reach 30 mm, brownish yellow in colour, often with white mottling laterally and elsewhere. The overall appearance gets darker with age. The dorsal cerata (up to 9 pairs) bear concentric rings of flattened pale tubercles. The frontal margin of the head, is somewhat dilated laterally.

This species is common all around the British Isles, in shallow water offshore, feeding upon hydroids (chiefly calyptoblasts such as *Nemertesia antennina* and *N. ramosa*). Elsewhere it has been recorded from Norway to the Mediterranean Sea, down to 200 m.

Family GONIODORIDIDAE

1. Dorsum without conspicuous ceratal processes **2**
Dorsum bearing elongated often brightly coloured ceratal processes . . **3**

2. Body colour milky white . . . . *Goniodoris nodosa* (p. 80)
Body colour brownish . . . . *Goniodoris castanea* (p. 80)

3. Ceratal processes limited to a pair at the level of rhinophores and a pair at the level of the gill circlet **4**
Ceratal processes more numerous **5**

4. Body white and yellow . . . . *Trapania maculata* (p. 88)
Body white . . . . *Trapania pallida* (p. 87)

5. Oral tentacles present . . . . *Ancula cristata* (p. 86)
Oral tentacles lacking **6**

6. Ceratal processes along mantle rim and in the central area of the back . **7**
No ceratal processes in the centre of the back **8**

7. Body colours white red and golden yellow . . . *Okenia elegans* (p. 82)
Body colours white suffused with pale pink . . . . *Okenia leachi* (p. 84)

8. Ceratal processes all along the mantle rim . . . *Okenia pulchella* (p. 85)
Ceratal processes present only alongside the gill circlet *Okenia aspersa* (p. 83)

Genus GONIODORIS Forbes and Goodsir, 1839

Goniodoris nodosa (Montagu, 1808) (Fig. 37)

Doris nodosa Montagu, 1808
Goniodoris emarginata Forbes, 1840

This species has a superficially delicate appearance but is in fact quite robust. It may reach 27 mm in extended length, coloured white with tinges on the dorsum of yellow and pink. There are patches of opaque white and also areas (notably around the bases of the rhinophores and behind the gills forming a false "pore") of especially translucent skin. The mantle edge is well developed but does not form an ample "skirt". A low keel runs down the middle of the back and on either side of this are small conical tubercles. The lamellate rhinophores are tinged with yellow. There are up to 13 pinnate gills around the anal papilla. The head bears a pair of lateral flattened oral tentacles. *G. nodosa* feeds upon encrusting polyzoans (such as *Alcyonidium*) when young, but the adults transfer their attention to compound ascidians especially *Botryllus schlosseri*, *Dendrodoa* and *Diplosoma*.

This is a common species on shores and in shallow waters (usually down to 20 m) all round the British Isles, although in other waters it has been reported down to 120 m. Outside Britain, *G. nodosa* has been recorded from Norway to Atlantic France, with an uncertain report from the Mediterranean Sea.

Goniodoris castanea (Alder and Hancock, 1845) (Fig. 38)

It is larger (up to 38 mm in length) than *G. nodosa* but is inconspicuous in the field with its red-brown body covered dorsally by white spots, the dorsum being tuberculate and ridged, as shown in Fig. 38. The mantle rim and a median dorsal crest-like ridge are especially notable. Rare specimens may be extremely pale all over. The lamellate rhinophores are brownish as are the gills. The latter are tripinnate and may be up to 9 in number, encircling the anal papilla. The sides of the head are produced to form spatulate oral tentacles. This is another species which is worldwide in distribution but when found is never abundant.

Around British coasts, this species lives in shallow water offshore, down to 25 m. Rarely it may be found in the inter-tidal zone. The prey is always a compound ascidian, especially *Botryllus* and *Botrylloides*. Outside Britain, *G. castanea* is known from Sweden to the Mediterranean Sea, Red Sea, New Zealand and Japan.

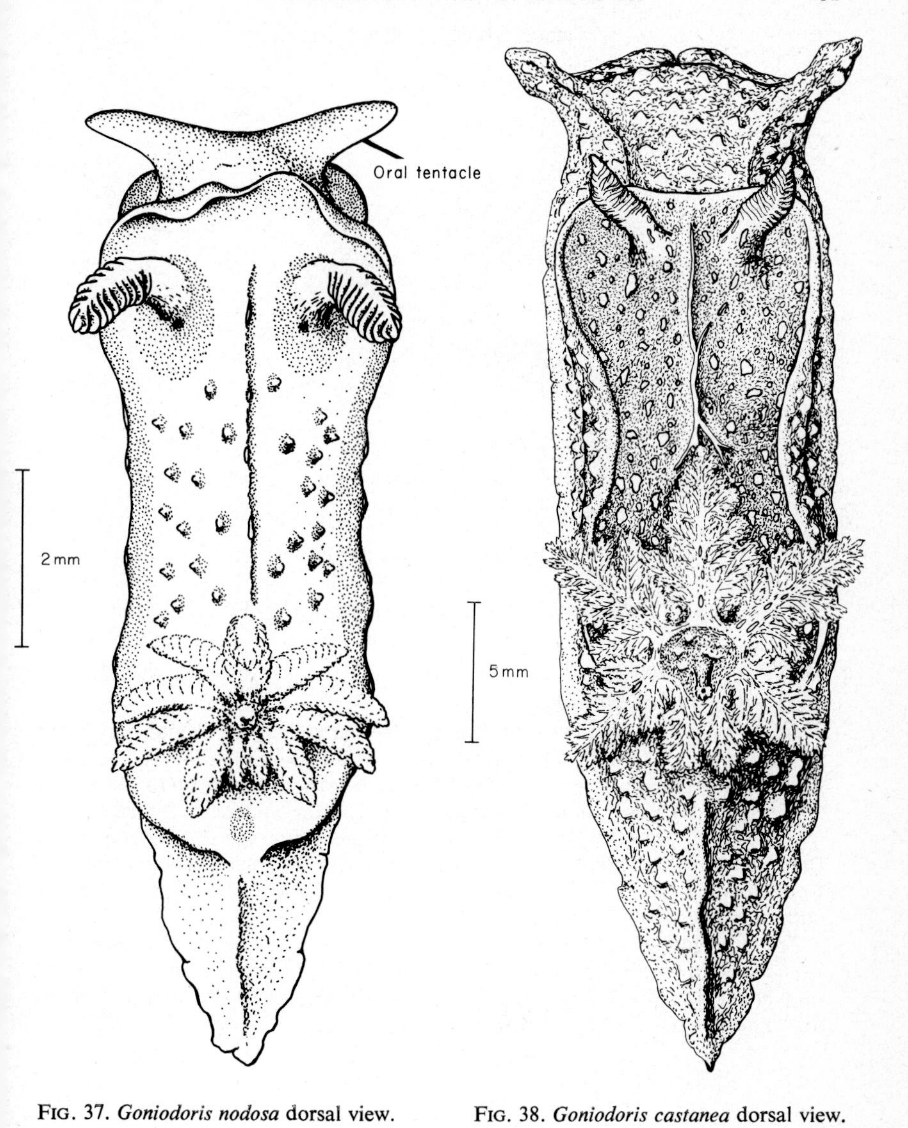

FIG. 37. *Goniodoris nodosa* dorsal view.

FIG. 38. *Goniodoris castanea* dorsal view.

Genus OKENIA Menke, 1830

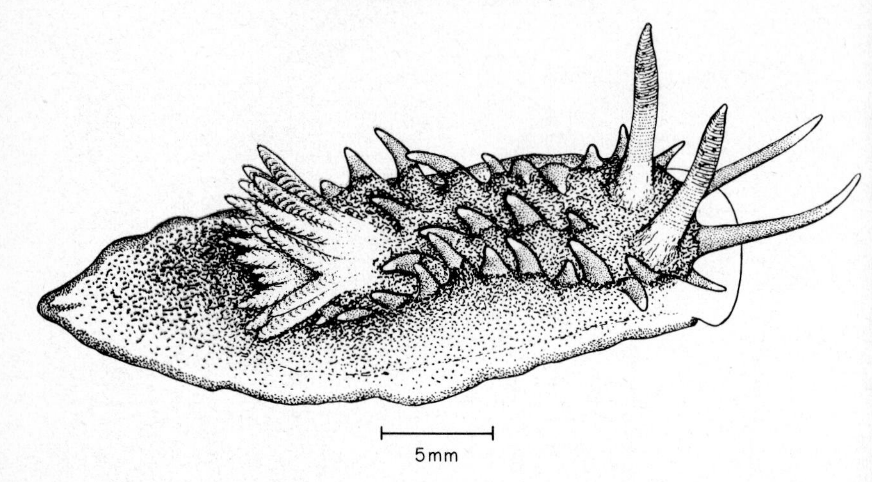

FIG. 39. *Okenia elegans* lateral view.

Okenia elegans (Leuckart, 1828) (Fig. 39)

Idalia elegans Leuckart, 1828
Idalia cirrigera Philippi, 1844

This is one of the most beautiful of the British nudibranchs, and may reach 80 mm in extended length. The body is white, suffused with pink to red. The dorsal processes are orange, with yellow or white tips. The mantle bears up to 35 tapering finger-like processes, along the pallial rim and in the middle of the back. Two (rarely 4) long anteriorly directed processes project from the skin in front of the rhinophores, which are themselves rosy coloured, with golden yellow lamellated tips. The gills (up to 21 in number) are similarly marked. The foot is bordered with orange-yellow.

This uncommon species feeds upon solitary ascidians (especially *Ciona* and *Molgula*) into which it burrows. It has been recorded from the Channel Islands, off Connemara, South Cornwall, and near Milford Haven, down to 30 m. Further distribution from the French Mediterranean coast.

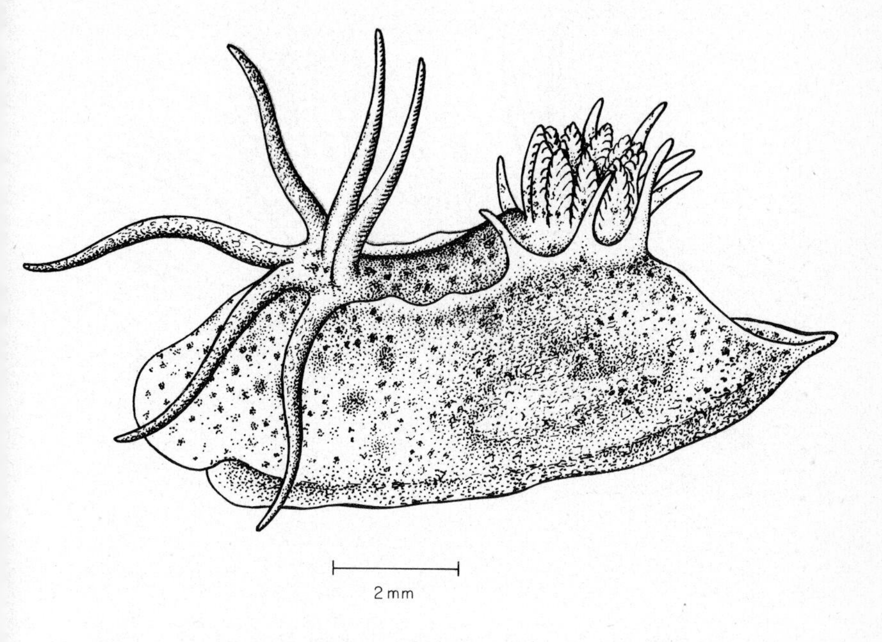

FIG. 40. *Okenia aspersa* lateral view (after Alder and Hancock (1845–55)).

Okenia aspersa (Alder and Hancock, 1845) (Fig. 40)

Idalia aspersa Alder and Hancock, 1845

Idalia inaequalis Forbes, 1851

The body may reach 22 mm in length, red or yellowish in colour, speckled with orange and brown. There are 2 pairs of long tapering anteriorly directed processes in front of the rhinophores and up to 4 pairs of shorter processes alongside the gills. The rhinophores are finely lamellate. Up to 12 simple pinnate gills encircle the anal papilla. The diet appears to consist of solitary ascidians, especially *Ascidiella* and *Molgula*.

This uncommon species has been recorded from Northumberland, the Irish Sea, and the Plymouth area, to 60 m. Elsewhere it is known from Norway to the French Biscay coast.

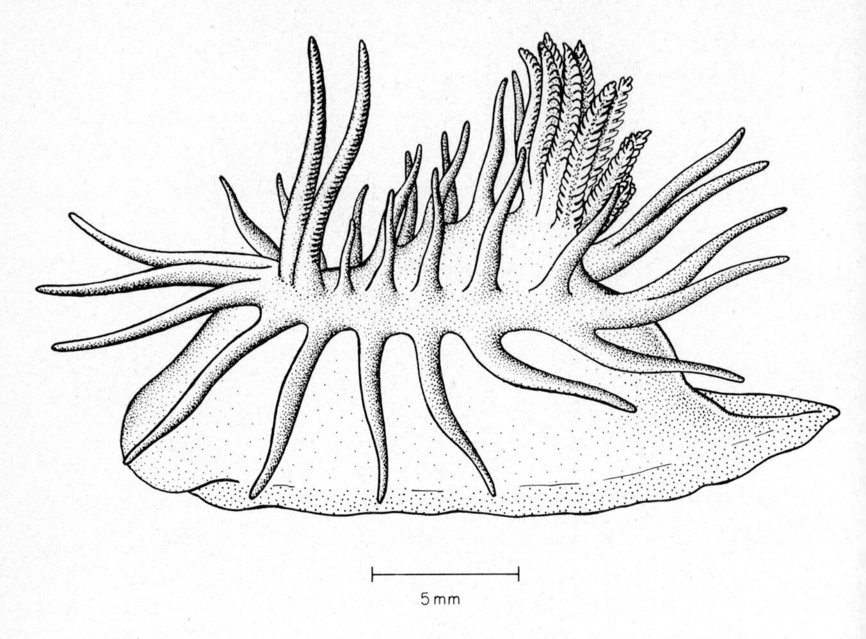

FIG. 41. *Okenia leachi* lateral view (after Alder and Hancock (1845–55)).

Okenia leachi (Alder and Hancock, 1854) (Fig. 41)

Idalia leachi Alder and Hancock, 1854

The body of this rare species may reach 25 mm in extended length. It is white, suffused with pink. The pallial rim bears a number of very long, tapering processes, 4 in front of the rhinophores, and 7 on either side of the body. The 2 most posterior processes are bifid. Inside the pallial rim are numerous shorter but still conspicuous processes. The rhinophores bear five lamellae. There are up to 11 gills in a circle around the anal papilla.

The diet of this species is unknown and more ecological information is urgently needed. Records exist, only from the last century, from the Hebrides, Durham, Connemara and Devon. All descriptions appear to be based on preserved material.

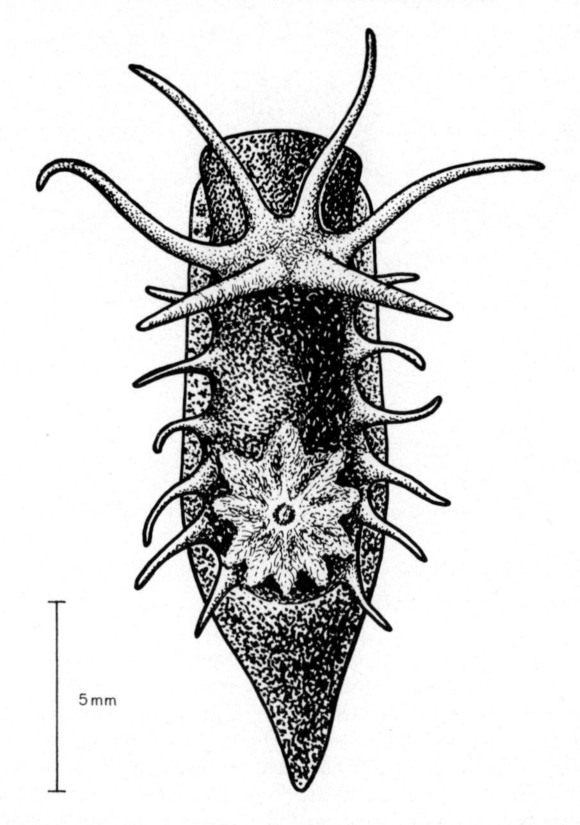

FIG. 42. *Okenia pulchella* dorsal view (after Odhner (1907)).

Okenia pulchella (Alder and Hancock, 1854) (Fig. 42)

Idalia pulchella Alder and Hancock, 1854
This is the rarest of the British species of *Okenia* and measures maximally only 19 mm. It is drab in colour and not easy to find. The body colour varies from pale pink to dull brown. The head is broad and especially well developed forming a shield-like projection in front of the rhinophores. The pallial rim bears laterally numerous finger-like processes, the most posterior of which may be bifid. Four long tapering processes project forwards from the mantle in front of the rhino-phoral tentacles. The rhinophores bear numerous fine lamellae. There are up to 11 gills in a circle around the anal papilla.

The diet is unkown and there is only one British record, dating from 1839 (Cornwall: *ca.* 40 m). Elsewhere it was recorded from Scandinavia over half a century ago.

Genus ANCULA Lovén, 1846

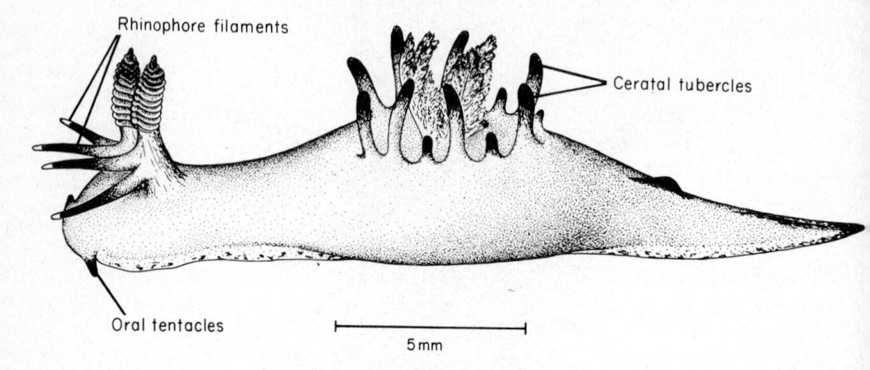

FIG. 43. *Ancula cristata* lateral view.

Ancula cristata (Alder, 1841) (Fig. 43)

Polycera cristata Alder, 1841

Miranda cristata Alder and Hancock, 1847

The body length may reach 33 mm, coloured white with brilliant yellow or orange tips. Rare specimens may be white all over. The shape is elongated and slender. At the level of the gills the mantle is produced on either side into strong, dorsally directed, ceratal tubercles (up to 6 or 7 on each side). There are 3 tripinnate gills. From the stalk of each rhinophore extends forwards a pair of stout filaments. The sides of the head are produced into short finger-like oral tentacles. It feeds upon compound ascidians such as *Botryllus* and *Botrylloides*.

This species is of sporadic occurrence in shallow waters all around the British Isles, occasionally on the sea shore. Further distribution from the White Sea to Brest, from Greenland and from eastern Canada, to 100 m.

Genus TRAPANIA Pruvot-Fol, 1931

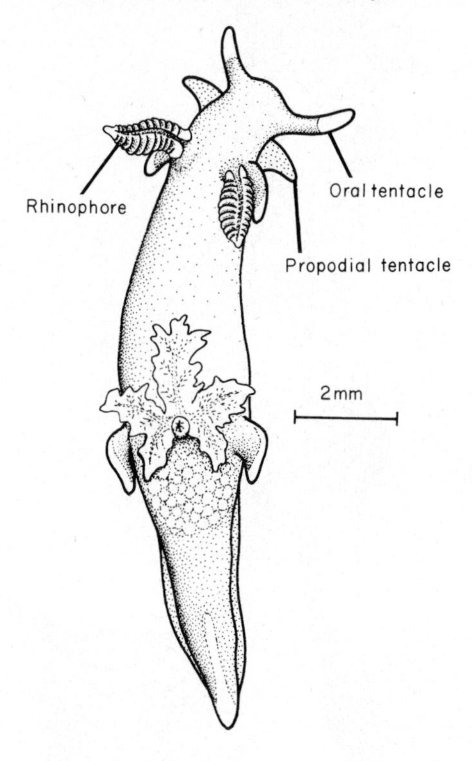

FIG. 44. *Trapania pallida* dorsal view (after Kress (1968)).

Trapania pallida Kress, 1968 (Fig. 44)

The body may reach 15 mm in extended length, uniformly cream in colour but with white patches on many parts of the dorsum and its appendages. The skin is smooth. A pair of conspicuous lateral processes projects from the mantle, one on either side of the gills. A similar process sticks out from the base of each rhinophore. There are 3 tripinnate gills anterior to the anal papilla. The oral tentacles are long and finger-like and there is also a conspicuous pair of propodial tentacles (Fig. 44).

This species has been found in Britain at Wembury, S. Devon, S. Cornwall, Lundy and N. Devon. Elsewhere it has been reported from Roscoff, France.

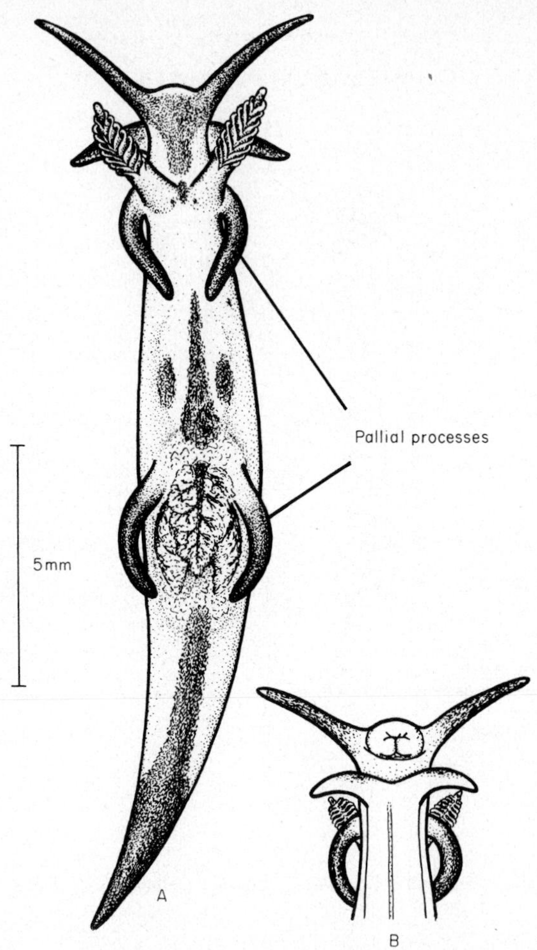

FIG. 45. *Trapania maculata:* A, dorsal view; B, ventral view of head.

Trapania maculata Haefelfinger, 1960 (Fig. 45)

This beautiful nudibranch attains a maximal length of 17 mm, white in basic colour but with numerous characteristic bright yellow markings. A pair of conspicuous curved lateral pallial processes project rearwards on either side of the gills and a similar process is attached close to the base of each lamellate rhinophore. Elongated pairs of oral and propodial tentacles are prominent. All these processes bear yellow pigment and there are yellow blotches on the epidermis of the dorsum as shown in Fig. 45.

Trapania maculata has been reported only once from British waters, on a boulder in 15 m off Portland Bill, where three specimens were found in June 1974 apparently feeding on tufts of an unidentified erect polyzoan. Elsewhere it is known only from the Mediterranean near Marseilles.

Family ONCHIDORIDIDAE

1. Low crenulated rhinophore sheaths present; mantle papillae tall and soft . . . . *Acanthodoris pilosa* (p. 90)
No rhinophore sheaths; mantle papillae rounded and stiff **2**

2. Dorsal mantle yellow-edged with red markings centrally *Onchidoris luteocincta* (p. 102)
Dorsal mantle no so marked **3**

3. Dorsal mantle blotched with brown **4**
Dorsal mantle uniformly white or yellow **9**

4. Gill pinnules up to 29 . . . . *Onchidoris bilamellata* (p. 96)
Gill pinnules never more than 12 . . . . **5**

5. Mantle tubercles very small, uniform, conical, spiculose (Fig. 53) *Onchidoris pusilla* (p. 104)
Mantle tubercles larger (Fig. 51A–D) **6**

6. Mantle tubercles rounded **7**
Mantle tubercles conical **8**

7. Area around each rhinophore conspicuously pigmented *Onchidoris sparsa* (p. 102)
No such areas evident . . . . *Onchidoris inconspicua* (p. 100)

8. Body extremely depressed . . . . *Onchidoris depressa* (p. 100)
Body not markedly depressed . . . . . . *Onchidoris oblonga* (p. 100)

9. Mantle tubercles very variable, some massive . . *Adalaria loveni* (p. 94)
Mantle tubercles varying but never massive **10**

10. Mantle tubercles rounded . . . . *Onchidoris muricata* (p. 98)
Mantle tubercles more elongated and tapering . *Adalaria proxima* (p. 92)

Genus ACANTHODORIS Gray, 1850

Acanthodoris pilosa (Müller, 1789) (Fig. 46)

Doris pilosa Müller, 1789
Doris stellata Gmelin, 1791
Doris flemingii Forbes, 1838
Doris sublaevis Thompson, 1840
Doris similis Alder and Hancock, 1842
Doris subquadrata Alder and Hancock, 1845

This soft-textured species may reach 50 mm in length but usually does not exceed 30 mm. The colour may vary from white through pale grey or brown to dark purplish brown or charcoal grey, usually uniformly distributed (not blotched or speckled) in any individual. Although speckled specimens have been recorded they are usually juveniles. The dorsal surface of the ample mantle is covered by soft, tall, conical, rather uniform tubercles. The rhinophores issue from low pallial sheaths which have denticulate rims. The rhinophore clubs when extended have a characteristic bend towards the rear. There are up to 9 voluminous tripinnate gills. Blunt, rounded oral tentacles are present. It feeds upon encrusting polyzoans, especially *Flustrella hispida* and species of *Alcyonidium*.

This common species occurs all around the British Isles. It is most often found on the shore but sublittoral records exist down to 80 m. Further distribution from the Arctic Ocean to the Mediterranean Sea, both seaboards of the U.S.A., and there are records which need confirmation from Australasia.

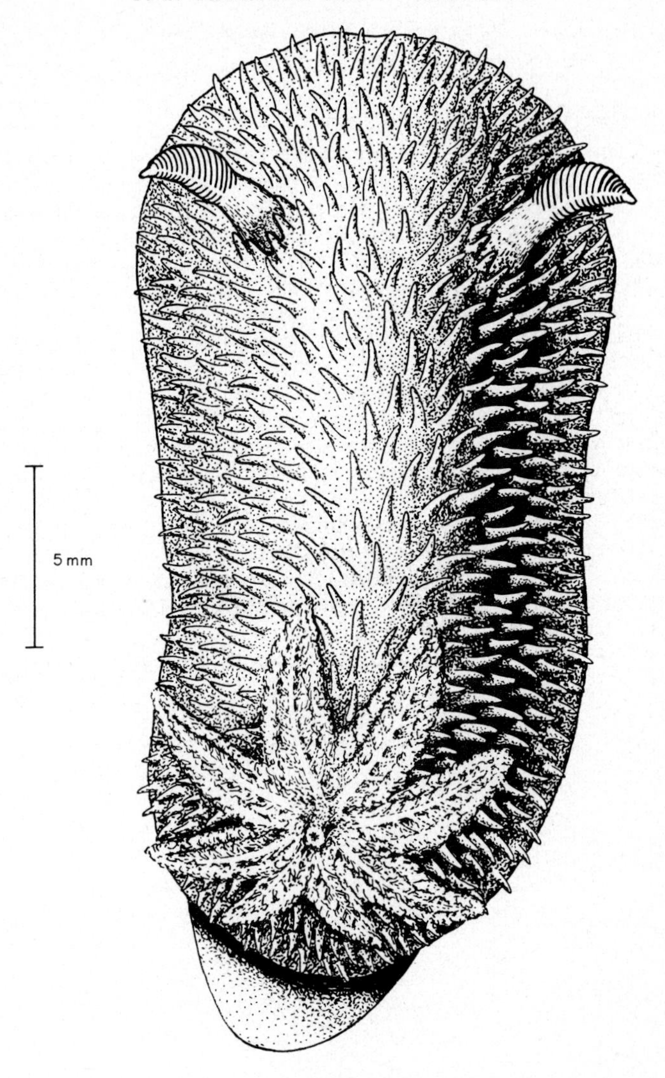

FIG. 46. *Acanthoris pilosa* dorsal view.

Genus ADALARIA Bergh, 1878

Adalaria proxima (Alder and Hancock, 1854) (Fig. 47)

Doris proxima Alder and Hancock, 1854

The body length may reach 17 mm in British specimens (up to 25 mm in other European localities), yellow in colour, sometimes very pale (white in extreme northern parts of the British Isles). The ample mantle bears abundant spiculose tubercles which are taller and more slender towards the periphery. The rhinophores and gills are often darker yellow than the rest of the body. Up to 12 pinnate gills are present surrounding the anal papilla. The head is dilated to form a semi-circular oral veil, lacking tentaculate processes.

This species is easily confused with *Onchidoris muricata* (Müller, 1776). *Adalaria proxima* has a generally deeper colour, the rhinophores are more bluntly terminated, and the digestive gland (visible through the translucent skin of the foot in ventral view) extends further forwards. But yellowish specimens of *O. muricata* can only be distinguished with certainty from pale *A. proxima* by examination of the internal anatomy, especially the radula. It feeds upon the polyzoan *Electra pilosa*.

Adalaria proxima may be locally common especially in the Menai Straits, but it has also been reported in other British coastal areas from Strangford Lough near Belfast to St. Andrews. It appears to be a boreo-arctic species and favours our northern coasts but there are records from the Bristol Channel and also one from the Plymouth area (by Garstang). Further distribution from East Greenland and the White Sea to the Baltic and Massachusetts, to 60 m.

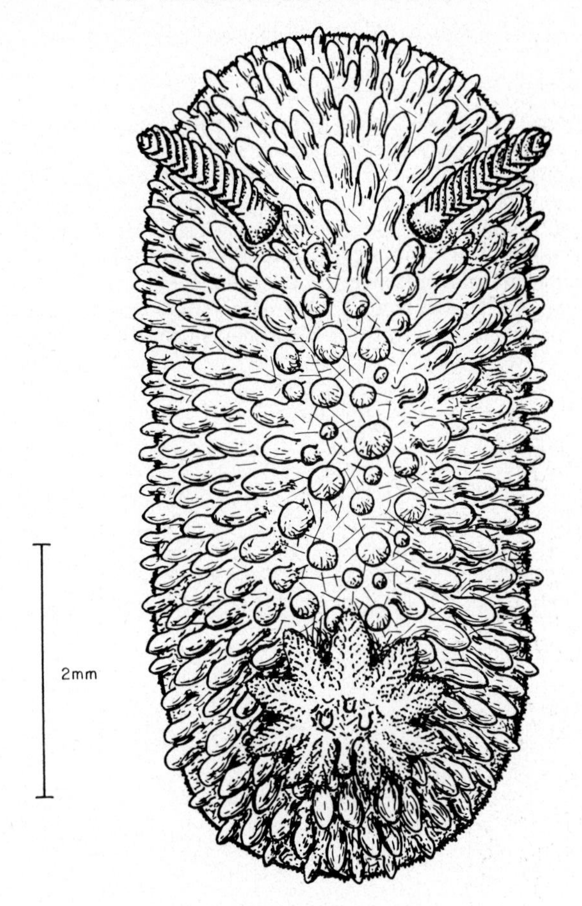

FIG. 47. *Adalaria proxima* dorsal view.

Adalaria loveni (Alder and Hancock, 1862) (Fig. 48)

Doris loveni Alder and Hancock, 1862

This rare species may reach 20 mm in length, pale yellow in colour. The ample mantle bears sparse large soft tubercles (up to 1·5 mm in height). There are up to 11 gills. The head is dilated to form a semi-circular oral veil, without tentacles.

Adalaria loveni could only be confused with a juvenile *Doris verrucosa* L., 1758 but can be immediately distinguished by the latter being cryptobranchiate (i.e., the gills can be retracted into a capacious gill cavity), whereas in *A. loveni* the gills are phanerobranchiate (separately contractile).

There is only one British record, from Bantry Bay, more than a century ago, although numerous records exist for Scandinavian coasts.

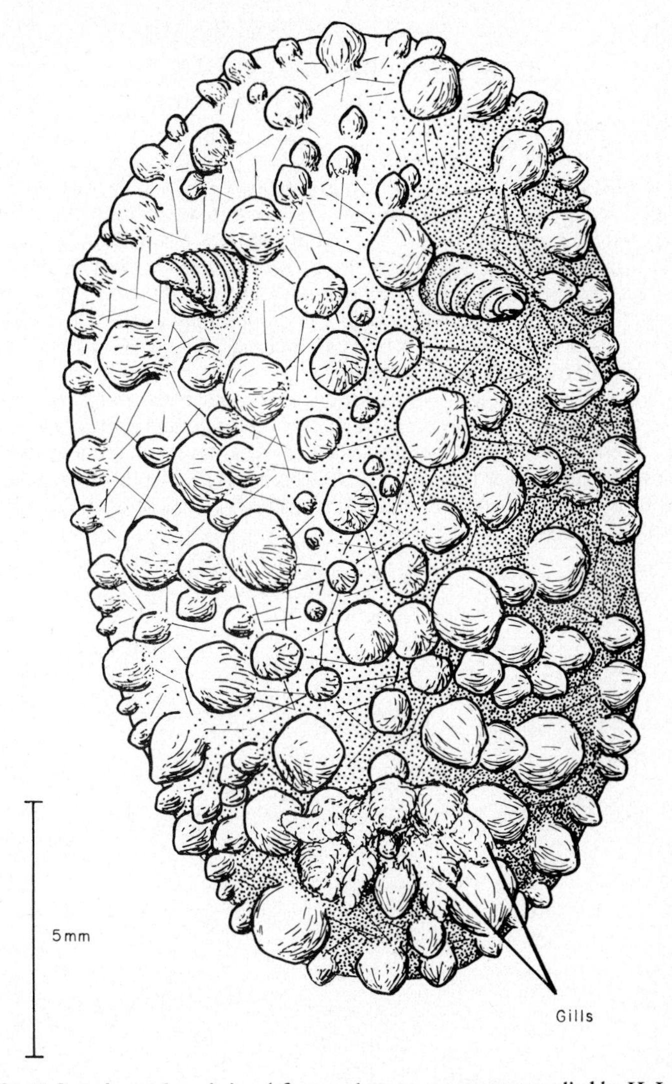

FIG. 48. *Adalaria loveni* dorsal view (after a colour transparency supplied by H. Lemche).

Genus ONCHIDORIS Blainville, 1816

Onchidoris bilamellata (L., 1767) (Fig. 49)

Doris bilamellata L., 1767
Doris fusca Müller, 1776

This common species may reach 40 mm in extended length, pale in ground colour but with blotchy brown markings on the dorsum (rare individuals, usually juvenile, may be pale all over). The ample mantle bears abundant spiculose club-like tubercles of various sizes. The brown pigment is usually absent from these tubercles. There may be up to 29 gills with, characteristically, a few tubercles within the branchial circlet. The pH of the mantle may drop to 1–3 on abrupt disturbance as defensive acid secretions are released. The head is dilated to form an oral veil without tentacular processes.

This species feeds upon a variety of species of acorn barnacles, but juveniles will also take encrusting polyzoans such as *Cryptosula* and *Umbonula*.

It has been recorded all around the British Isles, on the shore and in shallow water (usually not deeper than 20 m). Elsewhere it is known from Greenland and Iceland to the Atlantic coast of France, from both coasts of north America and from Japan, to 250 m.

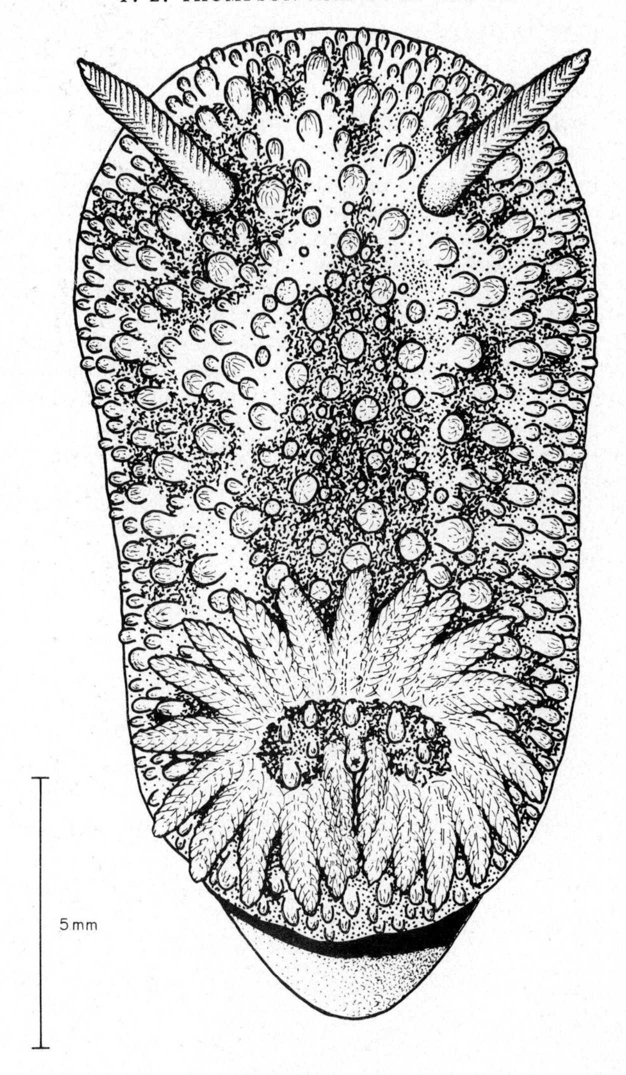

FIG. 49. *Onchidoris bilamellata dorsal view.*

Onchidoris muricata (Müller, 1776) (Fig. 50)

Doris muricata Müller, 1776

D. aspera Alder and Hancock, 1842

D. diaphana Alder and Hancock, 1845

This common species measures up to 14 mm in the field, although larger sizes have been recorded in pampered laboratory individuals. The colour is usually white but occasional pale yellow specimens may be found especially from extreme northern localities. The dorsal mantle bears abundant club-shaped spiculose tubercles. There are up to 11 pinnate gills around the anal papilla. The head is dilated to form an oral veil without tentacles.

This species is easily confused with *Adalaria proxima* (Alder and Hancock, 1854). *Onchidoris muricata* is generally paler in colour, the rhinophores are more pointed distally, and the digestive "gland" (visible through the translucent skin of the foot in ventral view) does not extend very far forwards.

Onchidoris muricata feeds upon a variety of encrusting polyzoans, especially *Electra pilosa* and *Membranipora membranacea.*

It occurs on the lower shore and in shallow offshore situations all around the British Isles, to 15 m. Further distribution from Greenland and the White Sea to Finisterre, from Alaska and from Nova Scotia, to 20 m.

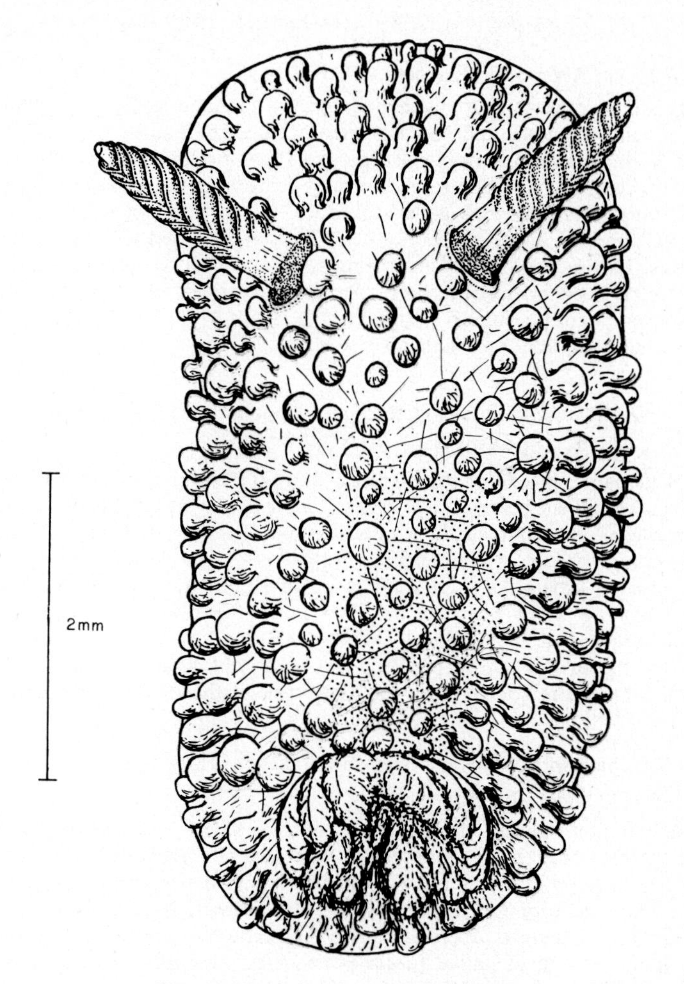

FIG. 50. *Onchidoris muricata dorsal view.*

Onchidoris depressa (Alder and Hancock, 1842) (Fig. 51A)

Doris depressa Alder and Hancock, 1842

This is an uncommon but distinctive species characterized by its small size (up to 9 mm in length) and very flattened profile. The mantle is pale brown in colour, with a pattern of bright orange spots tending, towards the mid-line, to possess a central black speck. A constant feature is the presence of evenly spaced, uniformly long, soft pallial tubercles. Up to 12 simple pinnate gills are present around the anal papilla. The head bears an oral veil without tentaculate processes. The prey consists of encrusting polyzoans, but no details are available.

Onchidoris depressa has been recorded from intertidal localities and shallow offshore waters in the English north-east, the Isle of Man, Devon and Cornwall. Elsewhere it is known from Banyuls-sur-Mer.

Onchidoris inconspicua (Alder and Hancock, 1851) (Fig. 51B)

Doris inconspicua Alder and Hancock, 1851

Despite its name, this species is no more (and no less) inconspicuous than other British species of the genus. *Onchidoris inconspicua* may attain a length of 12 mm and is always rather flattened in profile. The mantle is pale, with a tinge of purple, sprinkled with brown. A characteristic feature is the presence on the dorsum of small rounded tubercles. There may be up to 10 pinnate gills around the anal papilla. The oral veil lacks tentacles. This species feeds on encrusting polyzoans (e.g. *Cellaria sinuosa*) in shallow localities (to 50 m).

Around the British Isles it has been recorded only from Devon and the Irish Sea. Elsewhere it is known from the Skagerrak to the Bassin d'Arcachon.

Onchidoris oblonga (Alder and Hancock, 1845) (Fig. 51D)

Doris oblonga Alder and Hancock, 1845

This uncommon species may reach 12 mm in length, pale yellow-brown in colour with scattered brown and reddish blotches. The overall shape of the body when viewed from above is approximately oblong, hence the specific name. A more reliable feature is the presence on the dorsum of abundant, uniform, conical, stout, spiculose tubercles. Up to 7 pinnate gills are present around the anal papilla. A ring of pallial tubercles encircles these gills. This species preys upon encrusting polyzoans of the genus *Cellaria*.

There are very few reliable records of *O. oblonga*, and these are from scattered British localities all around our shores, down to 60 m. Further distribution from Sweden and the Øresund, with a doubtful record from Banyuls-sur-Mer.

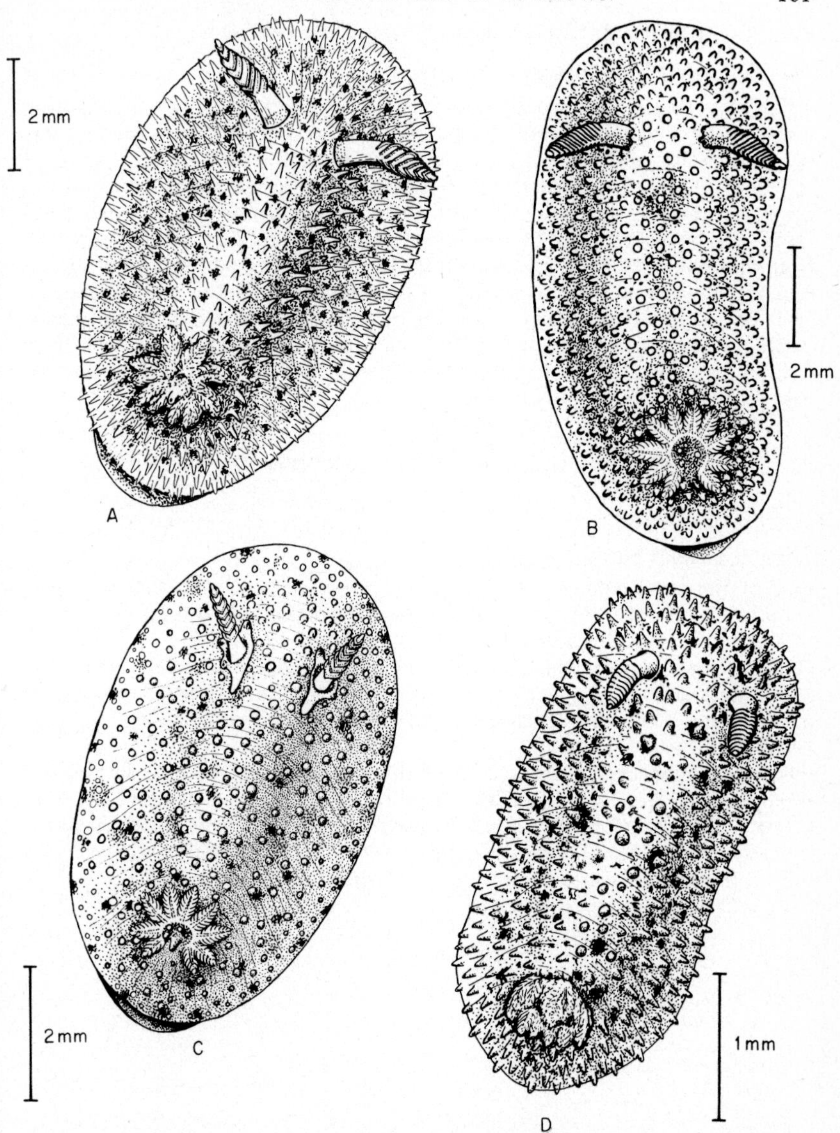

FIG. 51. *Onchidoris depressa:* A, dorsal view (partly after Alder and Hancock (1845–55)); *Onchidoris inconspicua:* B, dorsal view (after Alder and Hancock (1945–55)); *Onchidoris sparsa:* C, dorsal view (after Alder and Hancock (1845–55)); *Onchidoris oblonga:* D, dorsal view (after a colour transparency supplied by H. Lemche and partly after Alder and Hancock (1845–55)).

Onchidoris sparsa (Alder and Hancock, 1846) (Fig. 51C)

Doris sparsa Alder and Hancock, 1846

This exciusively British species does not appear to exceed 5 mm in extended length and often adopts a rounded shape when at rest, resembling a small colony of an encrusintg polyzoan. The mantle colour is usually pale brown with pink or red blotching. The back bears numerous small rounded spiculose tubercles which sometimes have a basal ring of black pigment together with a terminal black speck. The pigmentation of the mantle around the rhinophores is usually different from the remainder, sometimes darker, sometimes paler, and conspicuous pallial tubercles arise from these areas. Up to 10 simple pinnate gills may be present. The prey probably consists of encrusting polyzoans, but no details are available.

Only 5 specimens have been found, one from Northumberland, one from Orkney, two from the Plymouth area and one from the coast of Co. Galway. Further records are urgently needed.

Onchidoris luteocincta (M. Sars, 1870) (Fig. 52, plate I)

Doris luteocincta M. Sars, 1870

Doris beaumonti Farran, 1903

One of the most colourful British dorid nudibranchs, *O. luteocincta* reaches maximally 11 mm in length, white in ground colour but with lavish crimson blotching on the tuberculate dorsal mantle. Round the margin of the mantle skirt is a band of yellow which does not extend to the edge. The metapodium bears a well-marked keel dorsally. There may be up to 7 gills forming a circle around the anal papilla. The prey consists of encrusting polyzoans, including *Smittina*, according to Miller (1958).

Very few published records exist of this beautiful and delicate species and these were all from the Irish Sea. More recently abundant material has been brought up by aqualung divers in S. Devon, S. Cornwall, Lundy, Pembrokeshire, N. Devon and Belfast Lough, to 25 m. Outside the British Isles *O. luteocincta* is known from the Skagerrak and from the Mediterranean Sea to 50 m.

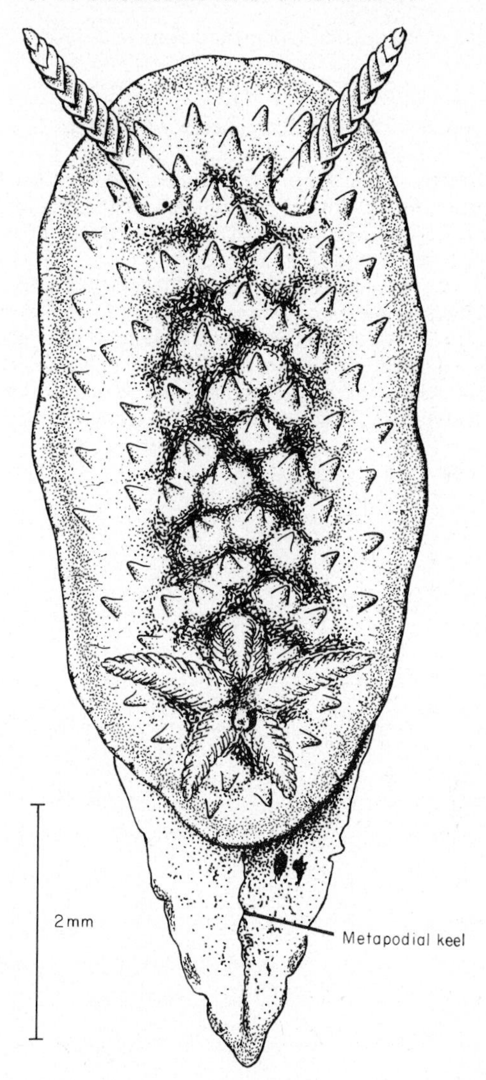

FIG. 52. *Onchidoris luteocincta* dorsal view.

Onchidoris pusilla (Alder and Hancock, 1945) (Fig. 53)

Doris pusilla Alder and Hancock, 1945

This is the darkest of the British species of *Onchidoris*, reaches a maximum length of only 9 mm and is inconspicuous. The dorsum is covered by dark brown spots, most strongly developed in the mid-line. The most reliable external feature is the presence all over the dorsum of abundant small, spiculose, conical tubercles. (These are smaller and less elongated than the pallial tubercles of *O. depressa* (Fig. 51A) or of *O. oblonga* (Fig. 51D)).

This species was regarded as uncommon until Miller (1958) recently obtained over 100 specimens in the Isle of Man. As with so many "rare" species, a precise knowledge of the individual's ecology was lacking so after Miller had discovered that *O. pusilla* fed on several species of encrusting polyzoans (*Escherella immersa*, *Escharoides coccineus*, *Microporella ciliata* and *Porella concinna*), he was able to find abundant material. Further distribution from Norway to Roscoff, down to 80 m.

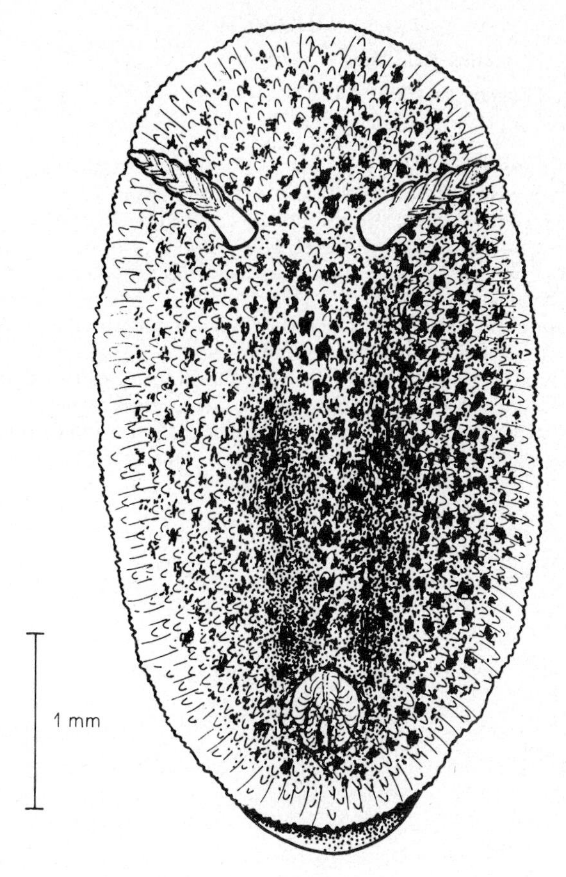

FIG. 53. *Onchidoris pusilla* dorsal view (after a colour transparency supplied by H. Lemche).

Family TRIOPHIDAE

Genus CRIMORA Alder and Hancock, 1862

Crimora papillata Alder and Hancock, 1862 (Fig. 54)

This uncommon nudibranch may reach 35 mm in length, usually white but sometimes pale yellow in ground colour, with numerous excrescences coloured yellow to orange as also are the rhinophores and gills. The pallial tubercles are forked in various ways and this is especially complex around the frontal margin of the head. Low but distinct rhinophore sheaths are evident. There are 3–5 tripinnate gills around the anus.

Until recently the only British record of this species was the original description in 1862 based on 2 individuals dredged among *Zostera* in Guernsey. Since that time it had been found only in the Mediterranean Sea (one specimen from Banyuls-sur-Mer and one from Morocco). Since 1972 aqualung divers have come across large numbers off the Pembrokeshire, Devon and Cornwall coasts in shallow water, down to 30 m, feeding upon *Flustra foliacea* and *Chartella papyracea*.

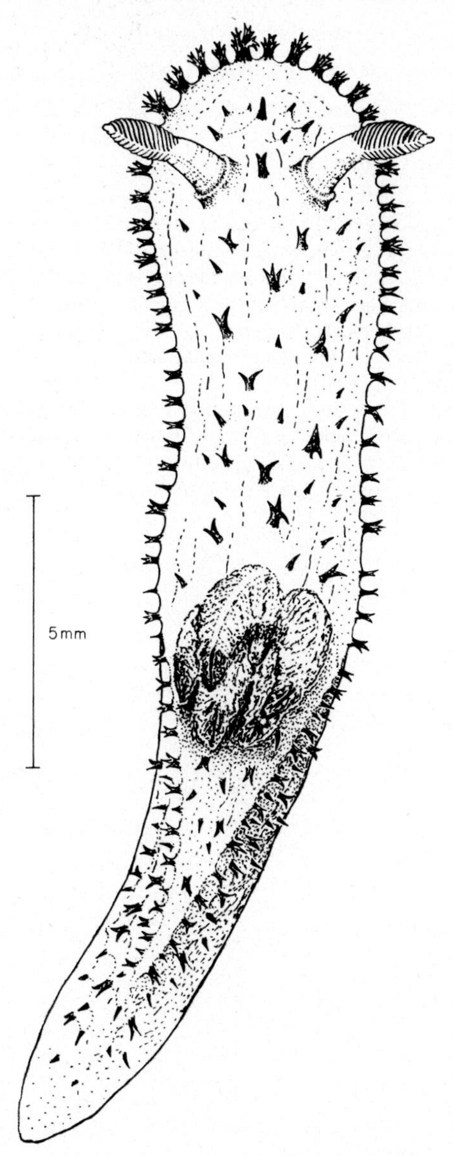

FIG. 54. *Crimora papillata* dorsal view.

Family NOTODORIDIDAE

Genus AEGIRES Lovén, 1844

Aegires punctilucens (Orbigny, 1837) (Fig. 55)

Polycera punctilucens Orbigny, 1837
Doris maura Forbes, 1840

The body length may reach 20 mm but is more usually less than 10 mm. The colour is predominantly brown, with numerous dark brown areas each of which contains a central brilliant blue spot. Juveniles may be paler, even white. Knobbed tubercles cover the dorsum and each of these bears an apical red spot. The smooth rhinophores issue from raised sheaths having tuberculate rims. There are 3 tripinnate gills anterior to the anal papilla. Short, rounded oral tentacles are present on each side of the head.

This species is extremely well camouflaged in its natural habit, among the encrusting polyzoans upon which it feeds. It has been recorded from unpolluted shallow waters all around the British Isles; elsewhere from Sweden to the Mediterranean Sea. A record from New Caledonia needs confirmation.

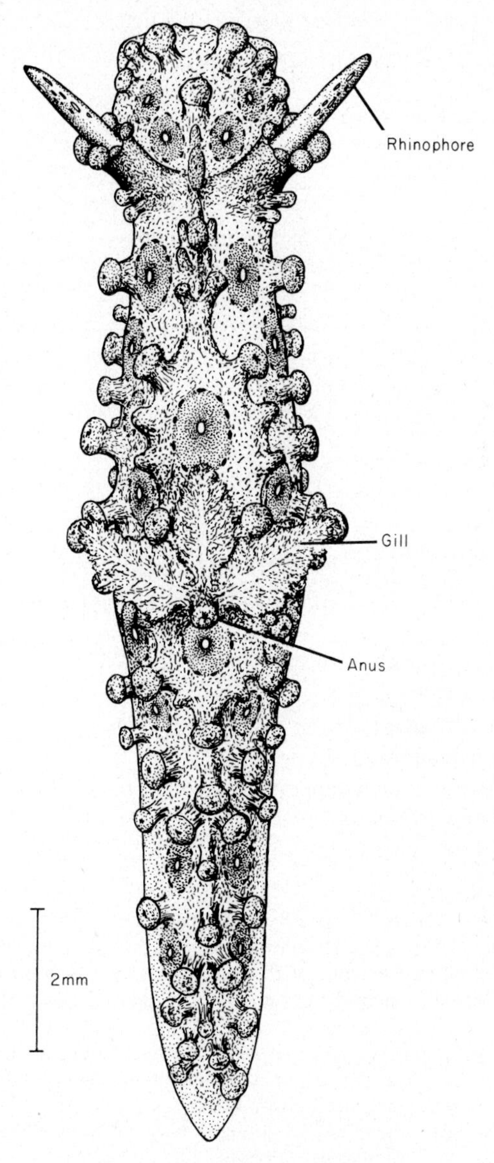

FIG. 55. *Aegires punctilucens* dorsal view.

Family POLYCERIDAE

1. Frontal margin bearing conspicuous finger-like processes 2
   Frontal margin tuberculate or smooth 4
2. Frontal processes smooth 3
   Frontal processes feathery . . . . *Limacia clavigera* (p. 120)
3. Up to 6 frontal processes . . . . *Polycera quadrilineata* (p. 110)
   From 6–12 frontal processes, usually more than 8 *Polycera faeroensis* (p. 112)
4. Mantle edge tuberculate; no rhinophore sheaths 5
   Mantle edge inconspicuous; conspicuous rhinophore sheaths present . 6
5. Body greenish yellow . . . . *Palio dubia* (p. 114)
   Body orange-red with blue spots . . . . *Greilada elegans* (p. 116)
6. Body white with greenish brown freckles . . . *Thecacera capitata* (p, 118)
   Body white with orange and black freckles . . *Thecacera pennigera* (p. 118)
   Body pinkish yellow marbled with green . . . *Thecacera virescens* (p. 118)

Genus POLYCERA Cuvier, 1817

Polycera quadrilineata (Müller, 1776) (Fig. 56)

Doris quadrilineata Müller, 1776
Policera lineatus Risso, 1826
Polycera ornata Orbigny, 1837
Polycera typica Thompson, 1840

This common species may attain a body length of up to 39 mm, coloured white with yellow or orange pigment located in blotches. There is usually a row of ovoid blotches down the mid-line of the back together with others on the flanks. Occasionally streaks and blotches of jet black may be present. Out of 100 individuals reported from the Isle of Man, 70 had such black markings. At the level of the gills the mantle is produced on each side into a strong, posteriorly directed, yellow or orange tipped ceratal tubercle. Similarly conspicuous tubercles project from the frontal margin of the head (usually 4 in number, as shown in Fig. 56). The rhinophores and the pinnate gills (up to 11 in number) are tipped with yellow.

Feeding upon *Membranipora membranacea* on laminarian fronds, this species may be found on the lower shore, but more usually in shallow offshore situations, down to 30 m. It may have been confused in the past with *P. faeroensis* but the two species are only superficially similar. *Polycera quadrilineata* has only 4 anterior processes and the pattern of orange body blotches is usually as shown in Fig. 56, while *P. faeroensis* has 8 or more anterior processes and sparser body blotches (Fig. 57). *Polycera quadrilinatea* occurs all round the British Isles and has also been recorded from Norway to the Mediterranean Sea, to 160 m.

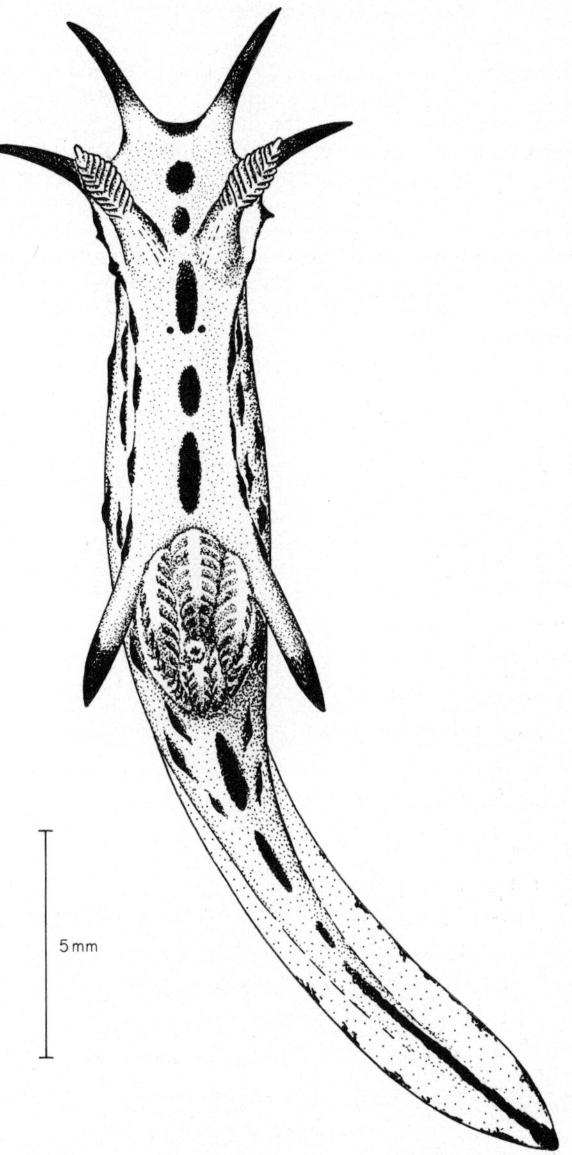

FIG. 56. *Polycera quadrilineata* dorsal view.

Polycera faeroensis Lemche, 1929 (Fig. 57)

This delicate sublittoral species (up to 43 mm in extended length) resembles superficially *P. quadrilineata* and because both species may occur in the same locality and sometimes feed upon the same prey, the polyzoan *Membranipora membranacea*, care must be exercised in their identification. The present species has at least 8 frontal processes and there are only rarely yellow blotches on the dorsum or flanks (compare Figs 56 and 57). The yellow pigment in *P. faeroensis* is usually paler than that of *P. quadrilineata*. The tendency towards black marking of the body evident in *P. quadrilineata* is a common feature also in *P. faeroensis*.

This species occurs commonly in shallow (to 25 m) waters off S. W. England, the Isle of Man, in Belfast Lough and Galway Bay. Elsewhere it is known only from Sweden and the Faroes.

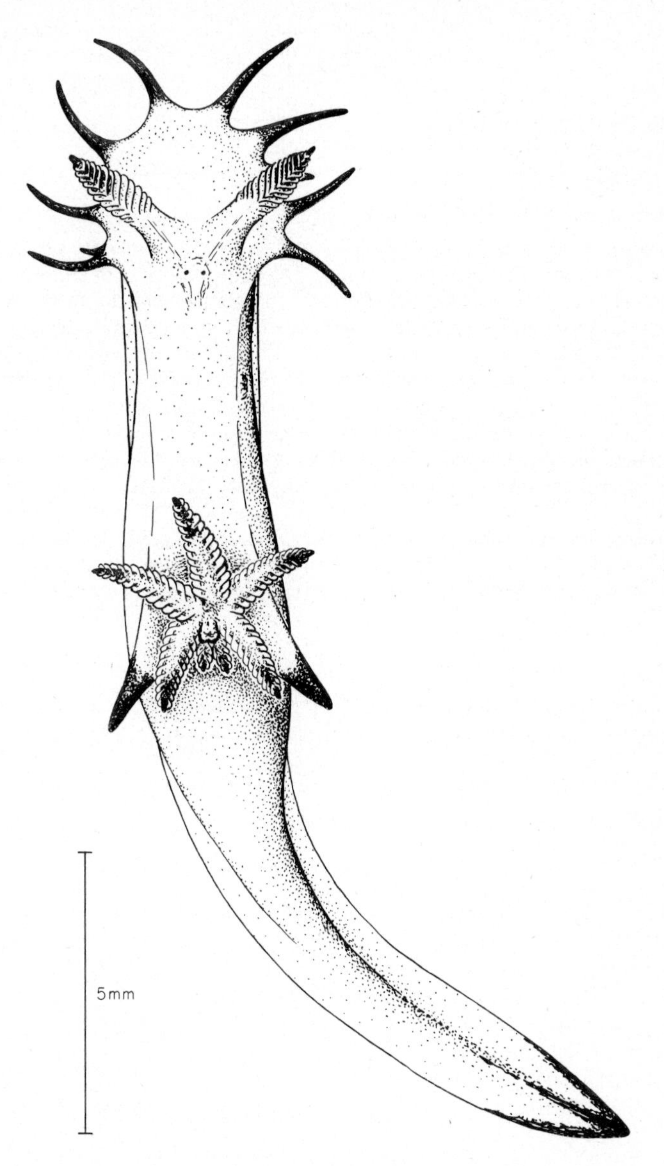

FIG. 57. *Polycera faeroensis* dorsal view.

Genus PALIO Gray, 1857

Palio dubia (M. Sars, 1829) (Fig. 58)

Polycera dubia M. Sars, 1829
Polycera lessoni Orbigny, 1837
Triopa nothus Johnston, 1838
Polycera ocellata Alder and Hancock, 1842

This inconspicuous species may reach 30 mm in length, but more usually does not exceed 20 mm. The ground colour is green or yellow with numerous low tubercles, always pale in colour. The rim of the mantle and the frontal margin are produced into similar tubercles, some of which may be more elongated and finger like. At the level of the gills the mantle rim forms on either side a large tubercle (lobulate in adults) which projects backwards. There are 3–5 tripinnate gills. A pair of conspicuous finger-like propodial tentacles is present.

This species varies considerably in the size of its tubercles and the deepness of the green coloration. It occurs in inter-tidal pools and down to considerable depths (100 m) feeding upon encrusting polyzoans such as *Schizoporella* and *Cryptosula*.

It is known to occur all around the British Isles; elsewhere it has been reported from the White Sea to the Mediterranean Sea, from Greenland and from E. Canada. A report from New Caledonia (Pacific Ocean) needs confirmation.

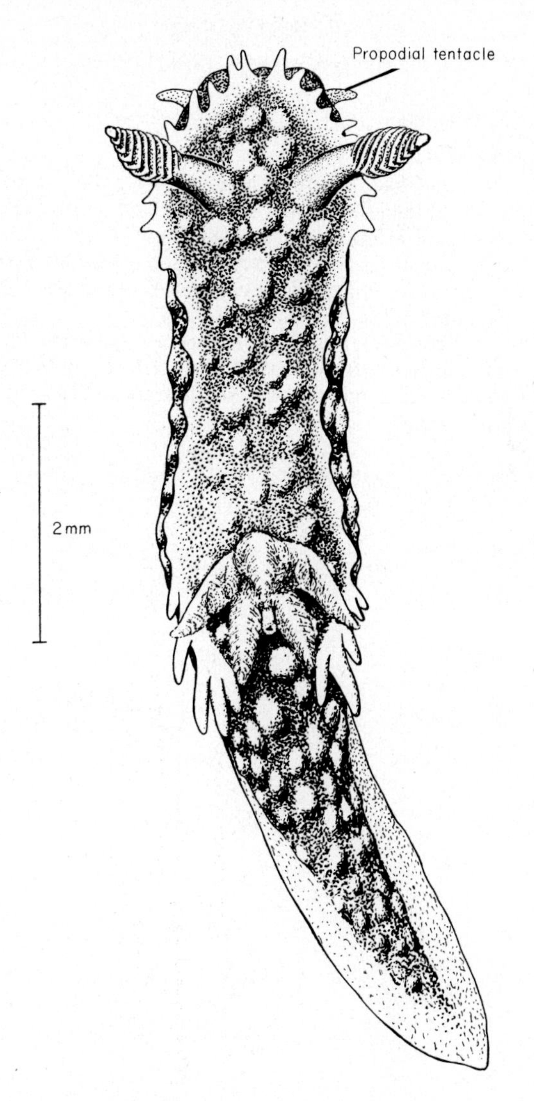

FIG. 58. *Palio dubia* dorsal view.

Genus GREILADA Bergh, 1894

Greilada elegans Bergh, 1894 (Fig. 59)

Polycera messinensis Odhner, 1941
?Polycera atlantica Pruvot-Fol, 1955

The length of the body may reach 40 mm, yellow to orange in colour with conspicuous brilliant blue blotches on the dorsum and flanks. These blue areas sometimes contain small crimson specks. The body is rather smooth, lacking the large anal tubercles characteristic of *Polycera* and of *Palio*. The frontal margin of the head bears up to 22 finger-like processes. There are 5–7 tripinnate gills in front of the anal papilla. The diet is unknown.

The only British records of this beautiful species are from Plymouth, from Lundy and from Pembrokeshire, in shallow water (down to 26 m). Elsewhere it has been reported from Arcachon and the Mediterranean Sea.

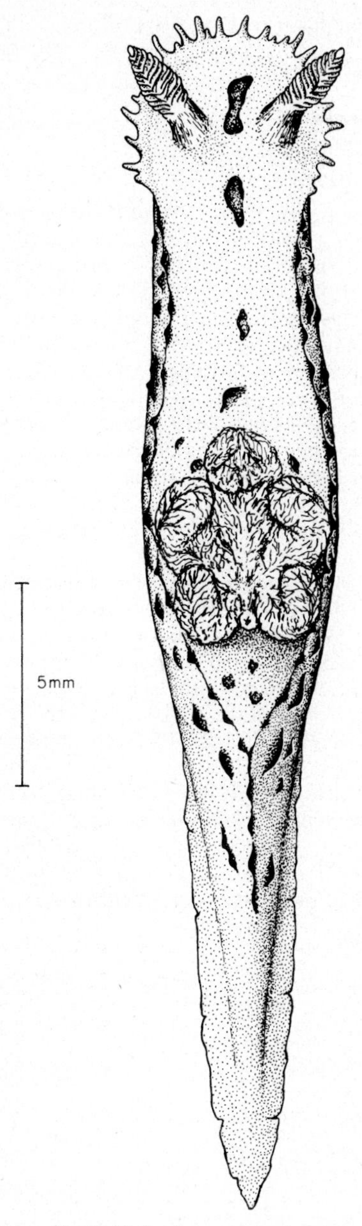

FIG. 59. *Greilada elegans* dorsal view.

Genus THECACERA Fleming, 1828

Thecacera pennigera (Montagu, 1815) (Fig. 60, Plate I)

Doris pennigera Montagu, 1815
?*Thecacera maculata* Eliot, 1905
?*Thecacera lamellata* Barnard, 1933

This dazzlingly marked nudibranch is, paradoxically, both rare and of worldwide distribution. It does not usually exceed 30 mm in extended length, with a ground colour of white, and numerous irregularly shaped orange blotches and jet-black and yellow spots over the entire dorsum. The black spots are always much smaller than the orange areas. The rhinophores issue from complex sheaths which are open mesially and rise to form substantial clubs behind. There are 3–5 bipinnate or tripinnate gills, forming a linked circlet around the anal papilla, and behind this circlet is a pair of massive club-like dorso-lateral processes. There are no oral tentacles but the front of the foot is produced into lateral points, the propodial tentacles. It is said to feed upon *Bugula*.

There are few British records of *T. pennigera*, from the southwestern coasts of England, Norfolk and Shetland. Further distribution from Arcachon and the Mediterranean Sea, Brazil, South Africa and Japan, always in shallow water, down to 20 m.

Two other species of *Thecacera* have been described from British coasts but they are so rare that their validity is hard to assess.

Thecacera capitata Alder and Hancock, 1854, is known from specimens dredged off north Cornwall in 1853. In length, these ranged up to 7 mm in length, in colour white with greenish brown freckled markings. There were 7 pinnate gills, tipped with orange, and just behind the anus a large club-like orange tipped tubercle on each side. The rhinophore sheaths were small and simple.

Finally, *Thecacera virescens* Forbes and Hanley, 1851 is again known only from Cornish material obtained well over a century ago. In length these animals reached 8 mm. The body was pinkish yellow, marbled with green. The rhinophore sheaths were small and simple. The most strange feature of this species is the presence of a row of blunt tubercles on either side of the gills.

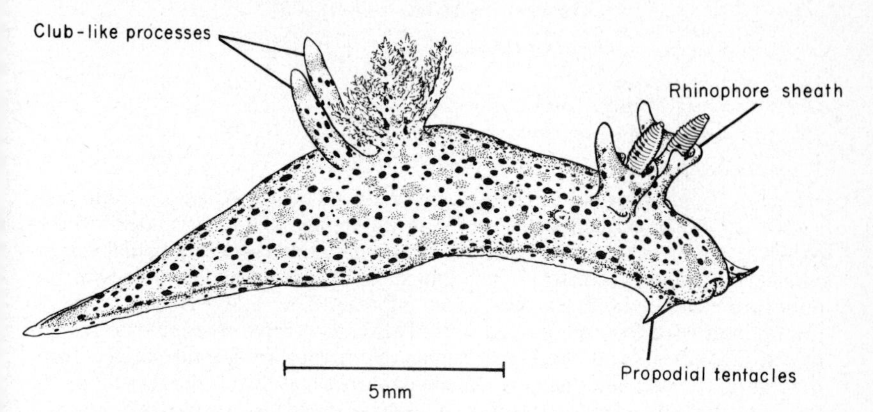

FIG. 60. *Thecacera pennigera* lateral view

Genus LIMACIA Müller, 1781

Limacia clavigera (Müller, 1776) (Fig. 61)

Doris clavigera Müller, 1776
Tergipes pulcher Johnston, 1834
Euplocamus plumosus Thompson, 1840

The body length may reach 18 mm (greater in the south Atlantic), white with yellow or orange extremities as shown in the figure. The mantle edge forms a paired series of substantial, elongated, finger-like processes, the pallial cerata, which are held erect. The frontal margin of the head bears similar processes and these are themselves beset with small papillae. The yellow-tipped lamellate rhinophores issue from smooth pockets. There are 3 (rarely 4) tripinnate yellow-tipped gills just in front of the anal papilla. Numerous low papillae project from the central dorsum and these are heavily pigmented, as also is the rear tip of the foot. The oral tentacles are conspicuous, elongated and characteristically grooved along the dorsal surface. It is known that many of these features are indetectable in juveniles.

This species has been recorded from clear shallow sublittoral areas (rarely from the sea shore) all round the British Isles, feeding upon a variety of encrusting polyzoans. Elsewhere it is known from Norway to the Mediterranean Sea, and from South Africa, down to 80 m.

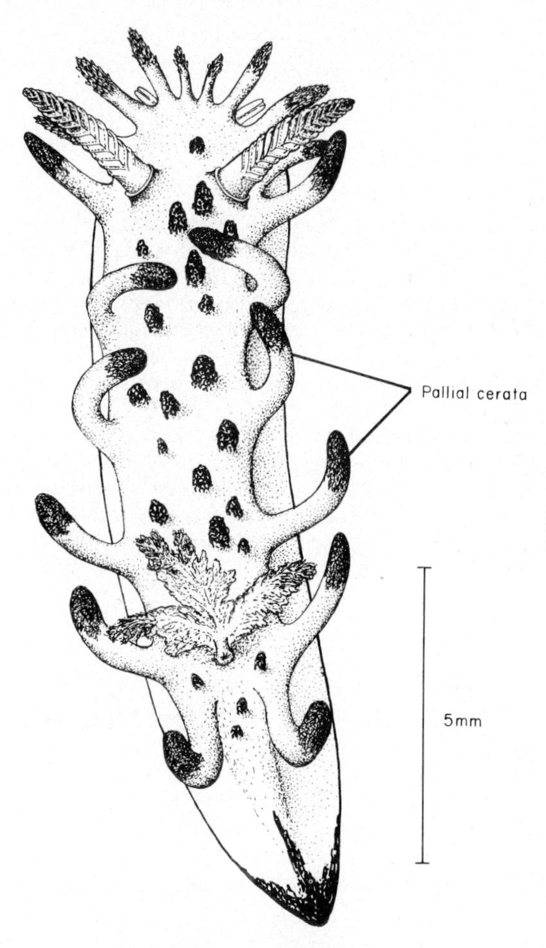

FIG. 61. *Limacia clavigera* dorsal view.

Family CADLINIDAE

Genus CADLINA Bergh, 1878

Cadlina laevis (L., 1767) (Fig. 62)

Doris laevis L., 1767

Doris obvelata Müller, 1776

Doris repanda Alder and Hancock, 1842

The body may reach 32 mm in length, flattened, delicate, translucent white in colour (rarely cream-yellow) with white or acid-yellow subepidermal glands in the ample but rather frail mantle skirt. The under-side of the mantle skirt exhibits a delicate white tracery of spiculose markings. The upper surface bears small conical soft tubercles. There are 5 tripinnate gills (rarely 7). The rhinophores issue from low, crenulate, pallial sheaths (see Fig. 62). The oral tentacles are short, broad and flattened. It has usually been found feeding on shallow water sponges.

This northern species (it is known to occur in water which is subzero throughout the year near Greenland) is commonest on the north-east of the British Isles but has been reported occasionally from the southwest of England and Wales, and from the east and south coasts of Ireland. Elsewhere it has been recorded down to 800 m from the Arctic Sea to the Mediterranean, and from both seaboards of the U.S.A.

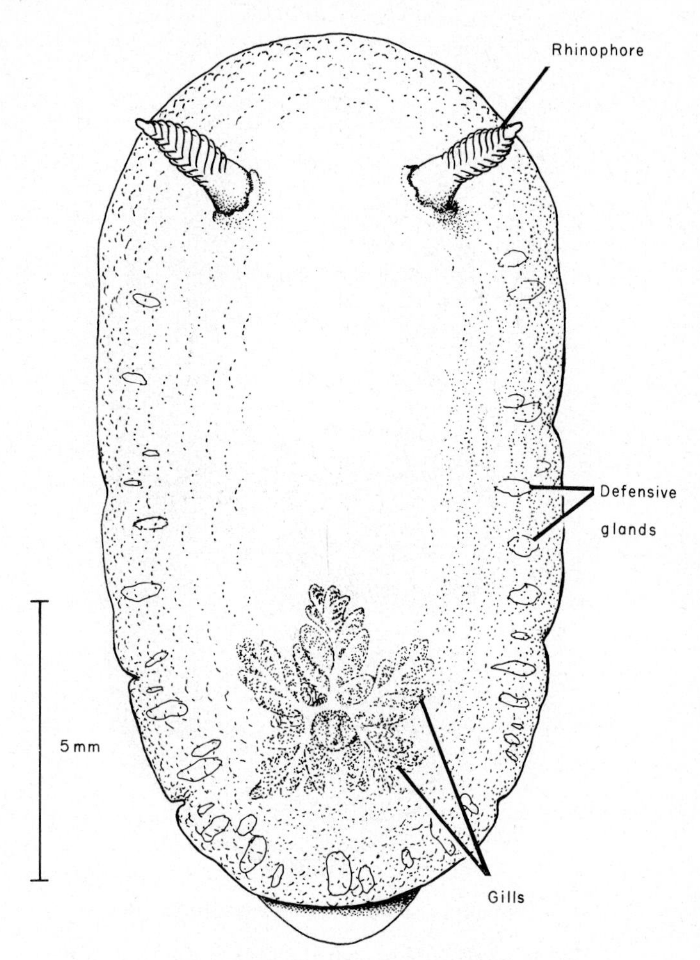

FIG. 62. *Cadlina laevis* dorsal view.

Family ALDISIDAE

Genus ALDISA Bergh, 1878

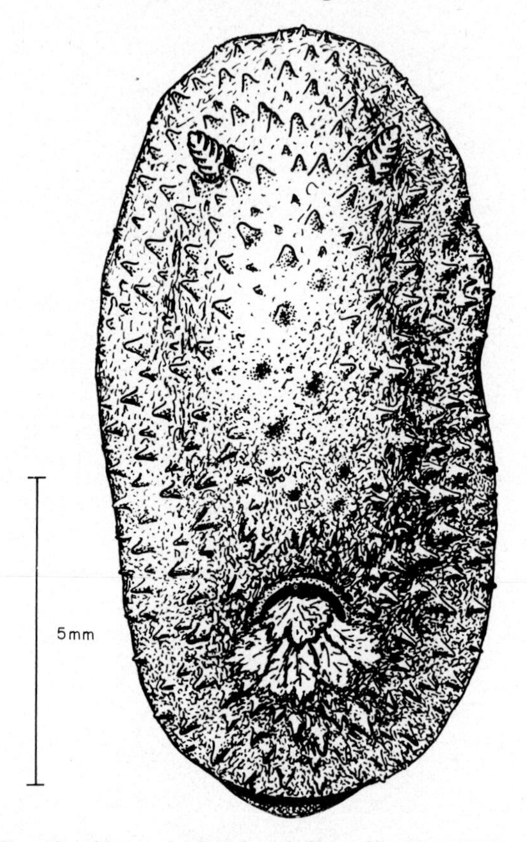

FIG. 63. *Aldisa zetlandica* dorsal view (after Sars (1878)).

Aldisa zetlandica (Alder and Hancock, 1854) (Fig. 63)

Doris zetlandica Alder and Hancock, 1854

This little known species may reach 35 mm in body length, grey-green in colour, with yellow gills and rhinophores. The dorsal surface of the ample mantle bears pointed conical tubercles. The margins of the rhinophores pits are each protected by 4–5 tubercles. There are 6–8 tripinnate gills. The oral tentacles are short and rounded.

The only British record of this species is from Shetland more than a century ago. Despite this, there can be no doubt that it was a well-founded species, and it has been recorded from Norway to the Azores, down to 1900 m.

Genus APORODORIS

Aporodoris millegrana (Alder and Hancock, 1854)

Doris millegrana (Alder and Hancock, 1854)

A more doubtful record is that of the strange *Aporodoris millegrana* (Alder and Hancock, 1854, as *Doris*) which has not been refound during the century that has elapsed since the type (2 specimens from Torbay) was described, and of which no live animals have been seen. An anatomical investigation would be necessary to be sure of identifying *A. millegrana*; the authors will be glad to help if a suspected *Aporodoris* is captured. Meanwhile this species is provisionally placed in the Aldisidae.

Family ROSTANGIDAE

Genus ROSTANGA

Rostanga rubra (Risso, 1818) (Fig. 64)

Doris rubra Risso, 1818

Doris coccinea Alder and Hancock, 1848

Rostanga rufescens Iredale and O'Donoghue, 1923

The body may attain a length of 15 mm, bright scarlet (rarely yellow) in colour with scattered black spots dorsally and a characteristic yellowish patch between the rhinophores. The ample mantle bears dorsally innumerable short blunt rather uniform tubercles. There are up to 10 simple pinnate gills. Digitiform oral tentacles project from the sides of the head. The lamellate rhinophores are yellowish in colour.

This conspicuous species feeds upon red sponges (such as *Microciona atrasanguinea*) in shallow waters on the western coasts of England and Wales. Elsewhere it has been recorded from Norway to Portugal and the Mediterranean Sea. A review of the world species of *Rostanga* has recently been published (Thompson, 1975).

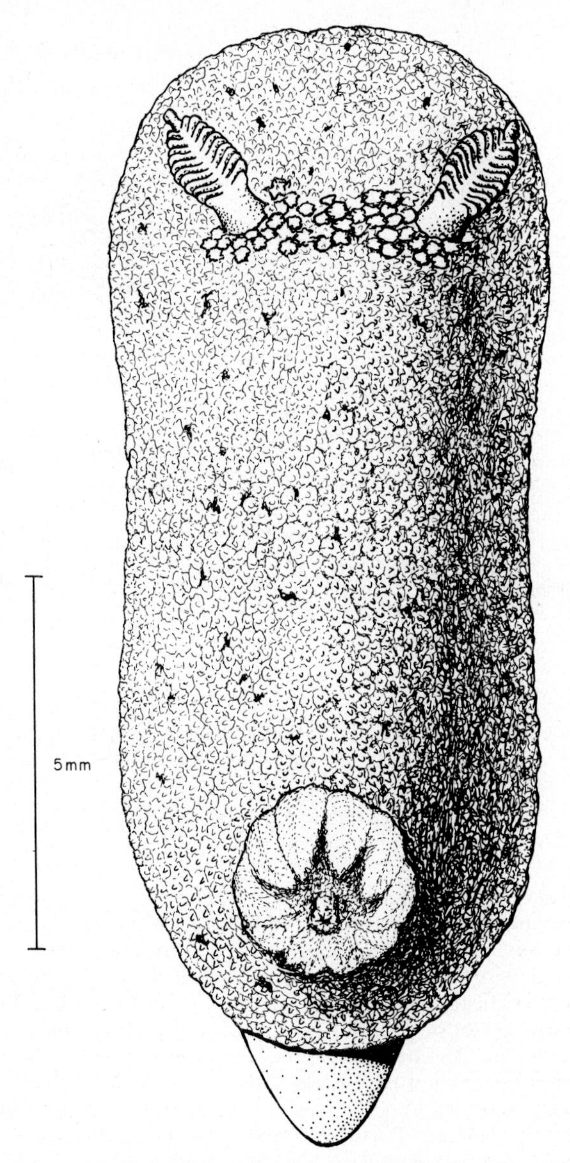

FIG. 64. *Rostanga rubra* dorsal view.

Family DORIDIDAE

Genus DORIS L., 1758

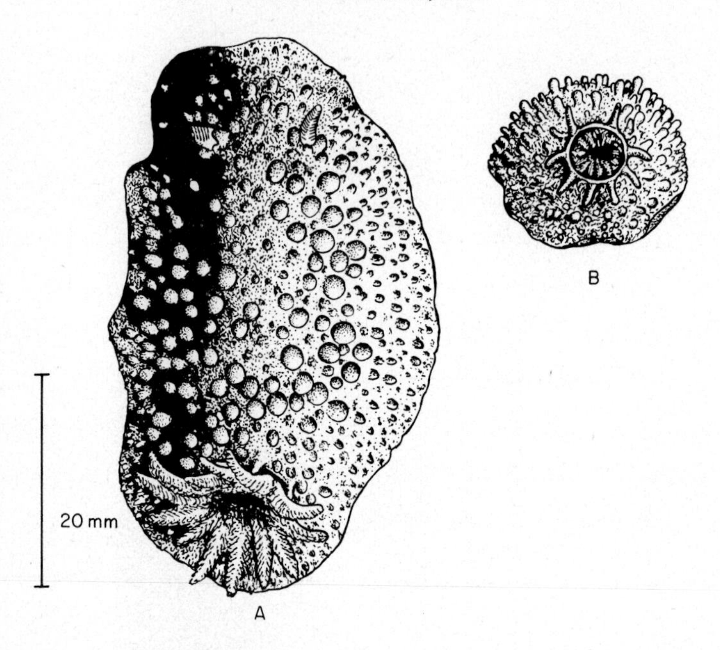

FIG. 65. *Doris verrucosa* (partly after Franz (1970): A, whole animal dorsal view; B, detail of branchial aperture with gills retracted.

Doris verrucosa L., 1758 (Fig. 65)

Doris derelicta Fischer, 1867

This species may reach 70 mm in extended length, grey or yellowish in colour, with an ample mantle covered by rounded tubercles, some of which may be very large (up to 3 mm in diameter), decreasing in size towards the mantle rim. The gill pocket and the rhinophore pockets are guarded by large tubercles which have a characteristically pedunculate shape. Up to 18 pinnate gills may be counted. A pair of short, grooved oral tentacles is present on the head. The pallial secretion is acidic. The natural prey is sponges but no details are available.

This dorid is uncommon on British shores; it has been recorded from the Firth of Clyde, W. Ireland and Devon. Elsewhere *D. verrucosa* is known from the French Biscay coast, and Portugal, from the Mediterranean Sea, South Africa, and both seaboards of the Americas, usually in shallow water.

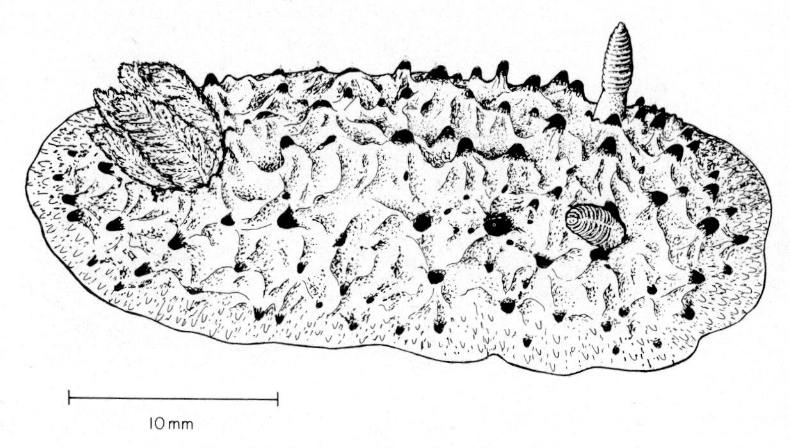

FIG. 66. *Doris maculata* dorso-lateral view.

Doris maculata Garstang, 1895 (Fig. 66)

Doridigitata sticta Iredale and O'Donoghue, 1923

The body of this beautiful but uncommon species may measure up to 45 mm in length, with a yellow ground colour topped off by purplish pigment on each mantle tubercle. These tubercles are up to 3 mm in diameter, inter-connected by a network of raised ridges. Especially strong tubercles guard the rhinophoral pockets. There may be up to 5 bipinnate or tripinnate gills. A pair of short, grooved oral tentacles is present on the head. Its diet is not known.

Doris maculata has been recorded on British coasts only from shallow water off Devon, Lundy and Pembrokeshire. Elsewhere it appears to have been recorded only from Marseilles.

Family ARCHIDORIDIDAE

Genus ARCHIDORIS Bergh, 1878

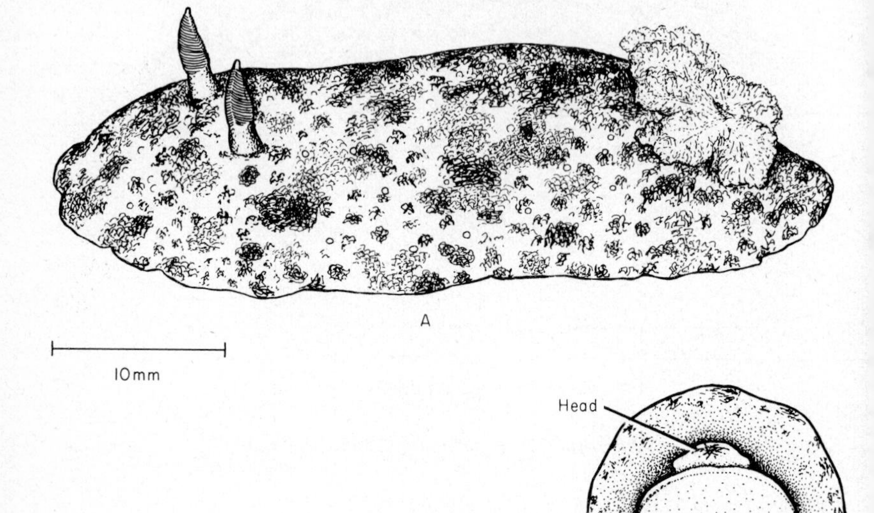

FIG. 67. *Archidoris pseudoargus:* A, whole animal dorso-lateral view; B, detail of head ventral view.

Archidoris pseudoargus (Rapp, 1827) (Fig. 67A and B)

Doris pseudoargus Rapp, 1827

Doris brittanica Johnston, 1838

Doris mera Alder and Hancock, 1844

Doris flammea Alder and Hancock, 1844

It is almost impossible to describe all the colours that may be found on the mantle of this the commonest and largest dorid nudibranch on British shores often referred to as the "sea lemon". It may reach 120 mm in length and the dorsum, covered by blunt tubercles of various sizes, bears blotchy markings of yellow, brown, pink, green and white pigments. The variety *flammea* is reddish in colour all over. There are 8–9 tripinnate gills which are paler than the mantle. The mouth bears vestigial oral tentacles, merely short tubercles one on either side of the head (Fig. 67B).

This shallow water and inter-tidal species feeds upon various sponges but chiefly depends on *Halichondria panicea*. It was at one time used for fish-bait in Shetland. It has been recorded all around the British Isles; further distribution from Norway to Portugal and the Mediterranean Sea, down to 300 m.

Genus ATAGEMA Gray, 1850

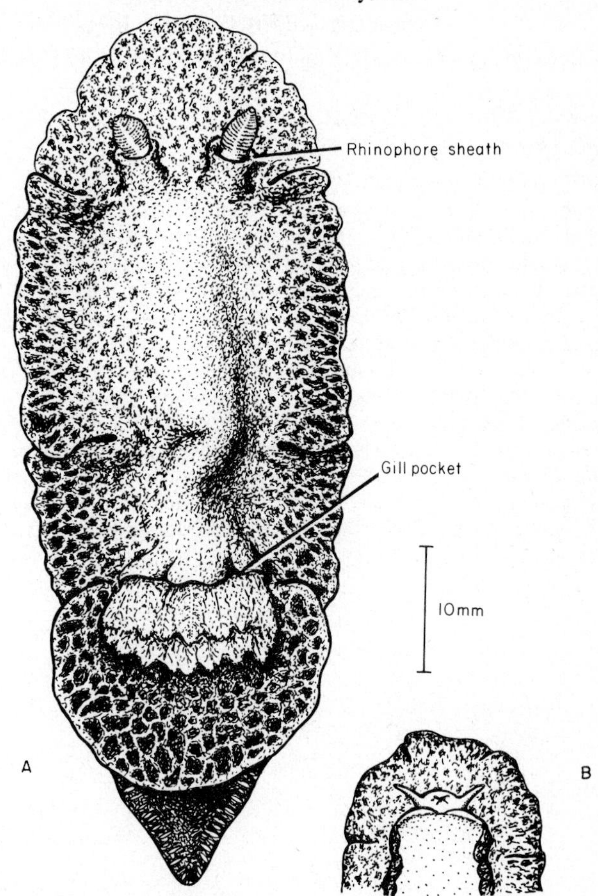

FIG. 68. *Atagema gibba:* A, whole animal dorsal view; B, detail of head ventral view.

Atagema gibba Pruvot-Fol, 1951 (Fig. 68A and B)

This rare species has been recorded up to 68 mm in length. The general aspect of the body is flattened but there is a conspicuous elevated zig-zag dorsal pallial ridge from between the rhinophores to the gills. The mantle rim is usually loosely crenulated. The dorsum feels stiff and velvety to the touch, covered with small spiny tubercles which form a net-like pattern. The papillae are pale while the intervening areas are rich brown. Trumpet-shaped rhinophores and a 5-lipped gill pocket further distinguish this species from any other British dorid nudibranch. Finger-like oral tentacles are present on the head (Fig. 68B).

Only one British locality is known for this distinctive species, near Porthkerris Point, Cornwall, under 12 m of water. *Atagema gibba* is known elsewhere only from Banyuls-sur-Mer, on the Mediterranean coast of France.

Family DISCODORIDIDAE

Genus DISCODORIS Bergh, 1877

Discodoris planata (Alder and Hancock, 1846) (Fig. 69A and B)

Doris planata Alder and Hancock, 1846
Doris testudinaria Alder and Hancock, 1862
Archidoris stellifera Vayssière, 1904

This uncommon species may reach 65 mm in extended length and bears a superficial resemblance to the common *Archidoris pseudoargus*. The mantle colour, however, is less colourful and variegated in *D. planata* and usually has a brown or purplish brown dominance. Numerous (up to 12) stellate pale areas form a loosely paired series on the dorsal mantle and these represent aggregations of acid gland openings. Their secretions are strongly acidic, contrary to the neutral mantle of *Archidoris*. The dorsum is covered by spiculose tubercles of generally smaller size than those of *Archidoris*. Finally, the head in *D. planata* bears conspicuous finger-like oral tentacles (Fig. 69B) lacking in *Archidoris*.

This shallow water species feeds on sponges, for instance *Hemimycale columella*, at various places all round the British Isles, but it is chiefly found on the southwest coasts. Elsewhere it has been recorded from Norway to the French Biscay coast (perhaps also from the Mediterranean Sea), always in shallow water.

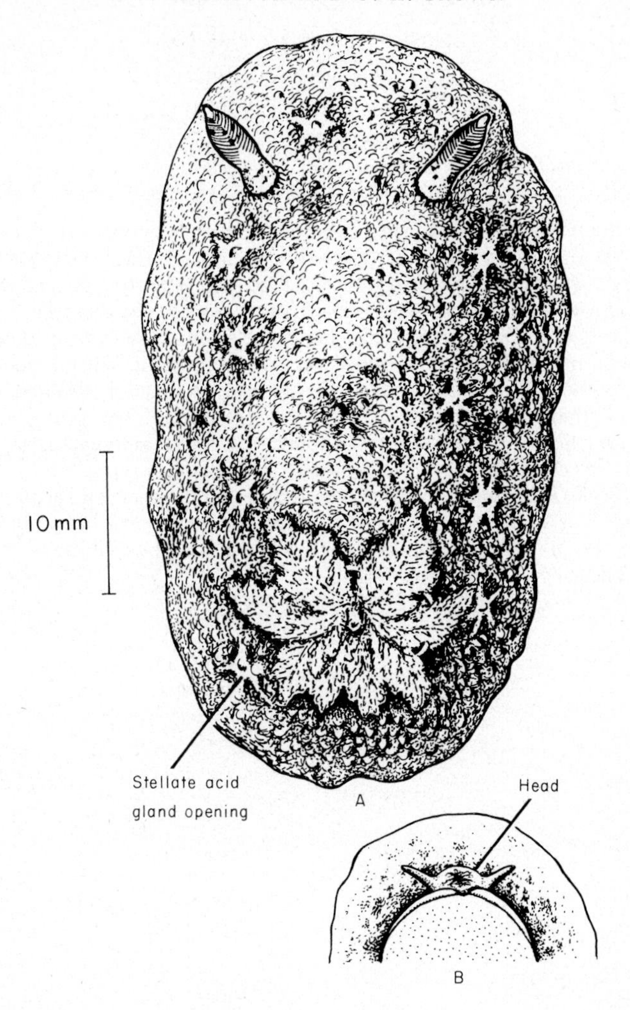

FIG. 69. *Discodoris pla ata:* A, whole animal dorsal view; B, detail of head ventral view.

Family KENTRODORIDIDAE

Genus JORUNNA Bergh, 1876

Jorunna tomentosa (Cuvier, 1804) (Fig. 70)

Doris tomentosa Cuvier, 1804

Doris johnstoni Alder and Hancock, 1845

This distinctive and common species has been recorded up to a length of 55 mm. The ample mantle is velvety and covered with small uniform spiculose tubercles, each of which has a characteristic retractile central projecting finger. The colour, is usually sandy brown but a fairly constant feature is the presence of a loosely paired series of dark brown blotches down the sides of the dorsum. The margins of the rhinophore pockets are slightly raised and serrated. The gill pocket, when the gills (up to 17 in number, tripinnate) are fully extended, is elevated and forms a short, cylindrical, vase-like base for the gill circlet. The oral tentacles are slender and finger-like. In Britain, *J. tomentosa* feeds upon encrusting siliceous sponges, especially *Halichondria panicea*.

This species is common on the shores and in shallow waters all around Britain; elsewhere it is known from Norway to the Mediterranean Sea and N. Africa down to 400 m. Records from the Pacific Ocean may be well-founded but need confirmation.

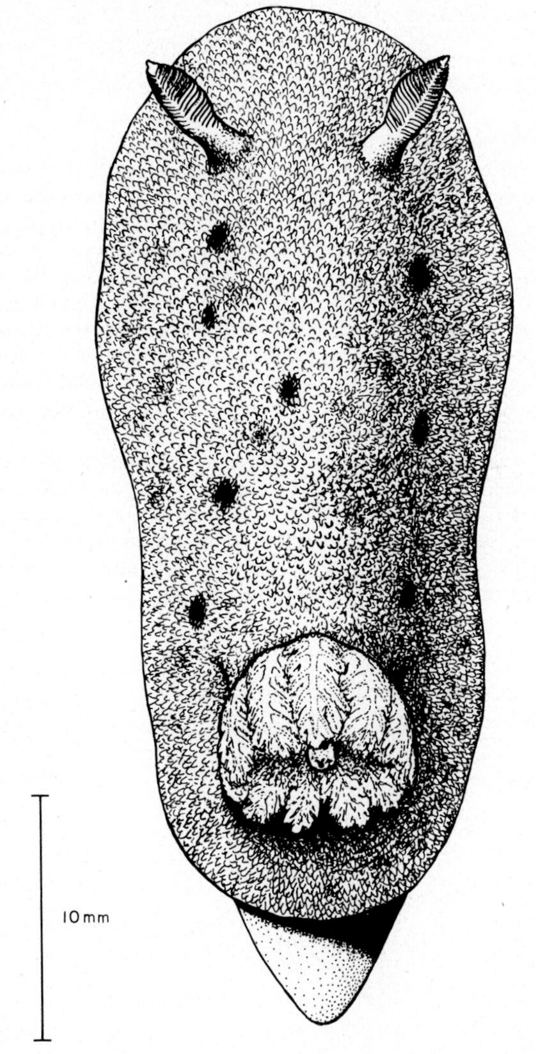

FIG. 70. *Jorunna tomentosa* dorsal view.

Family ARMINIDAE

Genus ARMINA Rafinesque, 1814

Armina loveni (Bergh, 1860) (Fig. 71A and B)

Pleurophyllidia loveni Bergh, 1860

This uncommon nudibranch may reach 40 mm in length. The ample mantle is brick-red in colour, with up to 50 prominent white longitudinal ridges. The frontal margin of the mantle is indented to protect the rhinophores which are united at the base. A wrinkled accessory caruncle appears to be situated just in front of the rhinophores but this and other morphological features need verification from live material.The mantle rim contains numerous large glands (brown in preserved animals). Under the rim is a symmetrical series of projections, the gills and the lateral lamellae, both of which are probably respiratory in function (Fig. 71A). The gills each consist of up to 20 longitudinal folds. The lateral lamellae on each side may be up to 30 in number. The head is large and flattened, produced laterally into pointed tentacles.

No finds of this species have been reported in print since 1910 and this description has been compiled from preserved material originally collected from the Irish Sea in 1949. *Armina loveni* seems to be associated with muddy sand below 40 m, perhaps feeding on coelenterates. It has been reported from northern coasts of Britain (from Durham to Cumberland) and also from Plymouth. Elsewhere, records exist from Norway and Sweden.

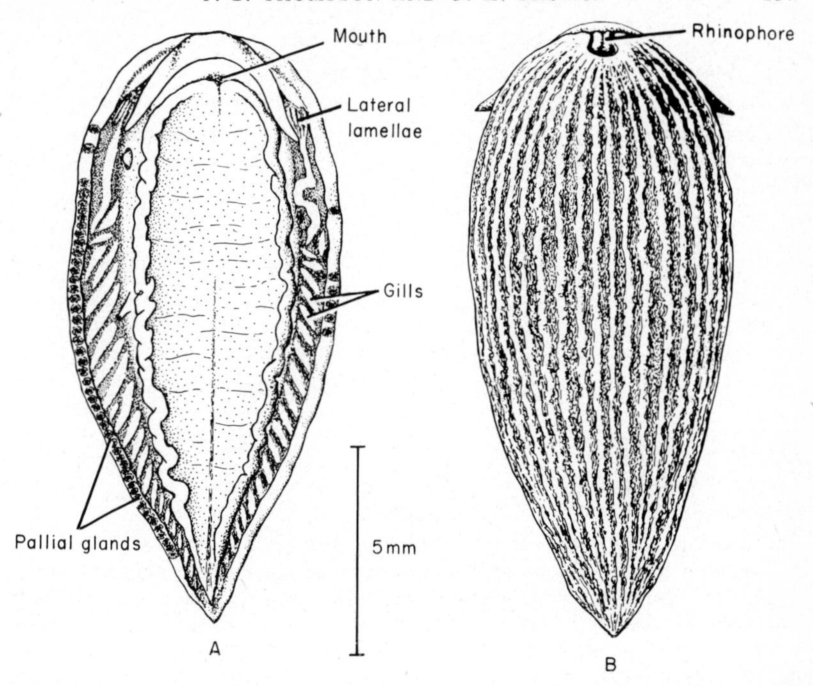

FIG. 71. *Armina loveni:* A, ventral view; B, dorsal view.

Family ANTIOPELLIDAE

1. Sensory caruncle present between rhinophore bases **2**
No such caruncle present . . . . *Proctonotus mucroniferus* (p. 142)

2. Ceratal processes smooth . . . . *Antiopella cristata* (p. 138)
Ceratal processes warty . . . . *Antiopella hyalina* (p. 140)

Genus ANTIOPELLA Hoyle, 1902

Antiopella cristata (Chiaje, 1841) (Fig. 72)

Eolis cristata Chiaje, 1841

Janolus cristatus (Chiaje, 1841)

Antiopa splendida Alder and Hancock, 1848

The extended length may be up to 75 mm, delicate, pale brown or cream in colour, with numerous conspicuous lateral cerata. These cerata are transparent, so that the central brown digestive gland can be seen running up the centre of each ceras and dividing at the tip. The tips of the cerata are peppered externally with white guanine pigment and have a bluish iridescent quality. Similar white pigment forms a line or blotches down the central dorsum, on the metapodium and on and around the lamellate rhinophores. The bases of the two rhinophores are united by a swollen wrinkled accessory caruncle. The anal papilla is postero-dorsal. The head bears short oral tentacles. This species feeds upon erect ectoproct Bryozoa such as *Bugula*.

It has been recorded from various localities off the English, Welsh, Scottish and Irish coasts. Elsewhere it has been recorded from the Netherlands and from the Mediterranean Sea.

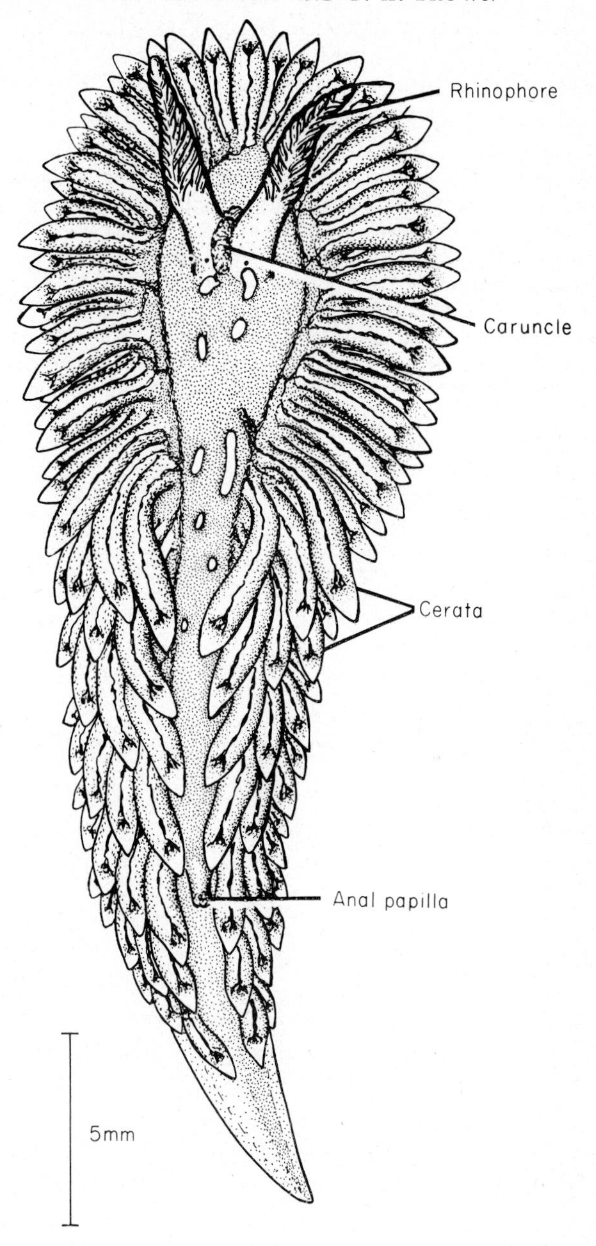

FIG. 72. *Antiopella cristata* dorsal view.

Antiopella hyalina (Alder and Hancock, 1854) (Fig. 73)

Antiopa hyalina Alder and Hancock, 1854
Janolus hyalinus (Alder and Hancock, 1854)

The length of this rare species may reach 20 mm, the body cream coloured with red-brown blotches. The numerous cerata are translucent white dotted with white and brown, exhibiting a brown digestive gland in the centre of each. These cerata are characteristically warty, as shown in Fig. 73. The bases of the lamellate rhinophores are united by a swollen wrinkled accessory caruncle. The anal papilla is postero-dorsal. The head bears short oral tentacles.

Antiopella hyalinus is one of our rarest species and it has been seen only 4 or 5 times, in scattered localities from the Isle of Man to Plymouth. There are no records from Scottish waters. Further distribution from Roscoff and Banyuls-sur-Mer in France. A recent Australian record awaits confirmation.

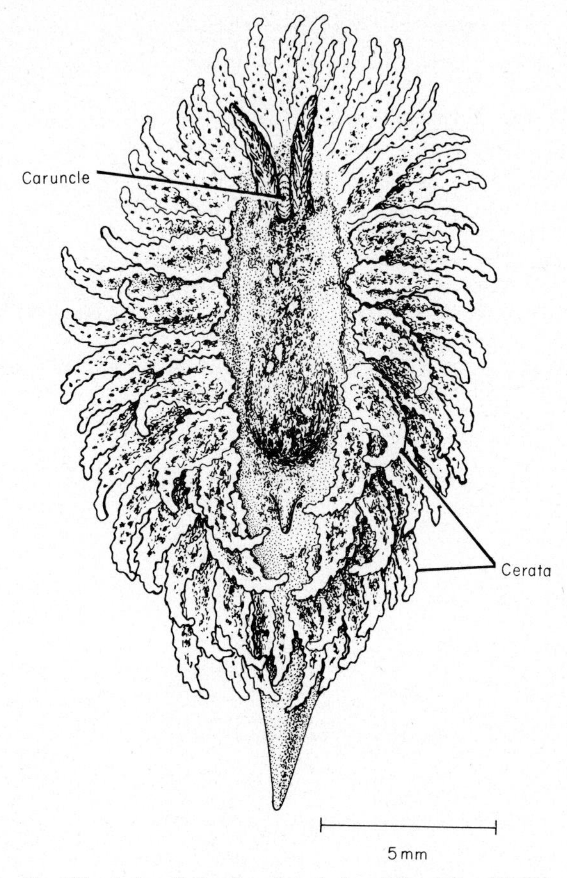

FIG. 73. *Antiopella hyalina* dorsal view (after Eliot (1910)).

Genus PROCTONOTUS Alder, 1844

Proctonotus mucroniferus (Alder and Hancock, 1844) (Fig. 74A and B)

Venilia mucronifera Alder and Hancock, 1844

Zephyrina pilosa Quatrefages, 1844

This is another extremely rare species, of which living material is urgently needed. The body may reach 11 mm in length, pale fawn marbled with brown. The numerous cerata are shining white, exhibiting a central yellow digestive "gland". The cerata are warty as shown in Fig. 74A. The lamellate rhinophores are separated at their bases and no accessory caruncle (see *Antiopella*) is present. The anal papilla is postero-dorsal. The head bears short oral tentacles (Fig. 74B). The diet is unknown.

This inconspicuous species has been seen only 6 or 7 times, always in northern localities, chiefly Irish. Outside Britain it has been recorded only from Roscoff in Brittany.

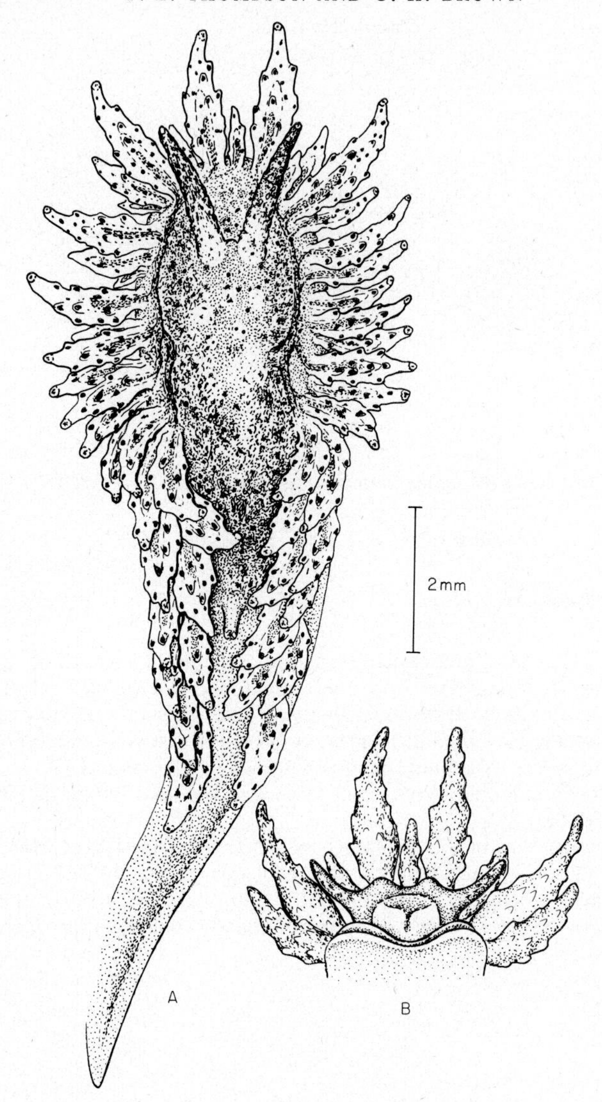

FIG. 74. *Proctonotus mucroniferus* (after Alder and Hancock (1845): A, dorsal view; B, head ventral view.

Family HEROIDAE

Genus HERO Alder and Hancock, 1855

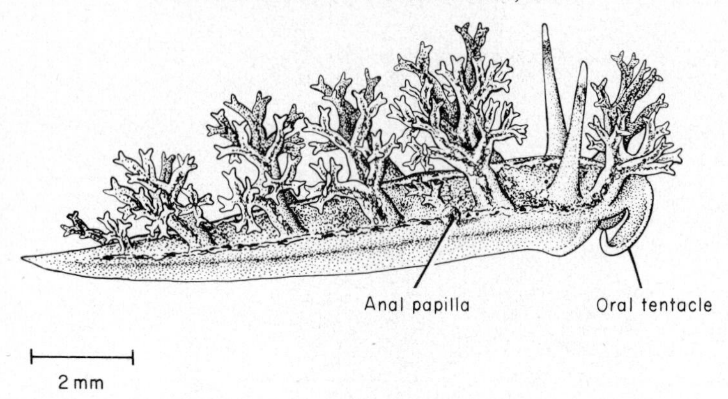

FIG. 75. *Hero formosa* lateral view (partly after Eliot (1910)).

Hero formosa (Lovén, 1841) (Fig. 75)

Cloelia formosa Lovén, 1841

Cloelia trilineata M. Sars, 1851

The body of this beautiful and delicate species may attain a length of 38 mm. In colour it is white or pink with 3 opaque white lines down the back. On each side of the back is a series of dichotomously divided yellow cerata (up to 8 cerata on either side); they each contain a grey or brown digestive gland branch. The smooth rhinophores lie behind the most anterior pair of cerata. The head bears a pair of greatly enlarged curved oral tentacles. The anal papilla is lateral, on the right side (Fig. 75).

This species may be locally common, for instance off the Isle of Man, feeding on both calyptoblastic and gymnoblastic hydroids (especially *Tubularia*), down to 60 m. There are records for the north, south, east and west of the British Isles, but elsewhere it is known only from Norway and the approaches to the Baltic Sea.

Family CORYPHELLIDAE

1. Body (excluding the cerata) purple . . . . *Coryphella pedata* (p. 145)
Body (excluding the cerata) white **2**

2. Opaque white median line present . . . . *Coryphella lineata* (p. 146)
No such lines . . . . *Coryphella verrucosa* (p. 148)

Genus CORYPHELLA Gray, 1850

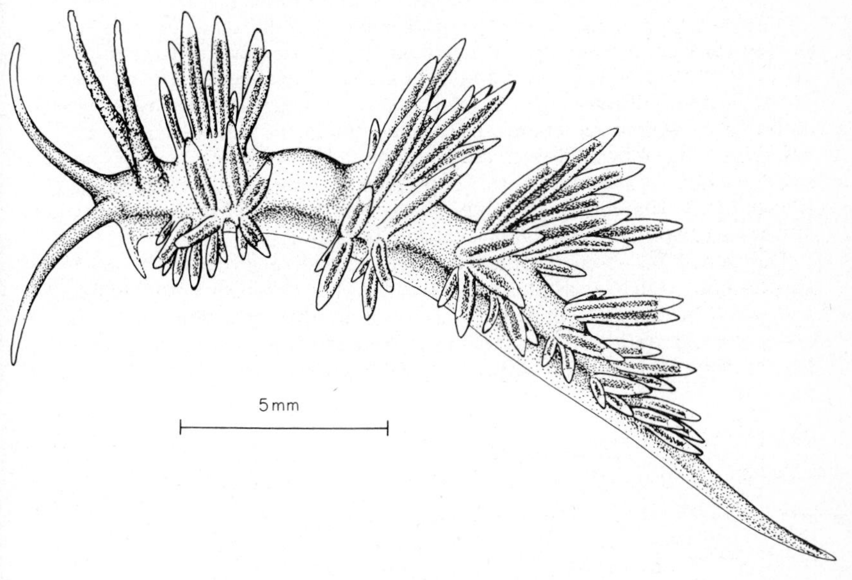

FIG. 76. *Coryphella pedata* dorso/lateral view.

Coryphella pedata (Montagu, 1815) (Fig. 76)

Doris pedata Montagu, 1815

Eolis landsburgii Alder and Hancock, 1846

This slender and beautiful species, which may reach 40 mm in length, has a violet-coloured body which immediately sets it apart from any other British nudibranch (although in the Mediterranean Sea there may be the possibility of confusion with *Flabellina affinis* (Gmelin, 1791), which possesses lamellate not wrinkled rhinophores). The cerata are arranged in up to 6 latero-dorsal clusters, and each ceras contains an orange-red digestive lobule. The tips of the oral tentacles, rhinophores and cerata are white.

This sublittoral species can be locally abundant, feeding upon both gymnoblastic (chiefly *Tubularia* and *Garveia*) and calyptoblastic hydroids (e.g. *Abietinaria* and *Hydrallmania*), all around the coasts of Britain, to 40 m. Further distribution from Norway to the Mediterranean Sea.

Coryphella lineata (Lovén, 1846) (Fig. 77)

Aeolis lineata Lovén, 1846
Aeolis argenteolineata Costa, 1866

This may attain a length of 40 mm, translucent white in ground colour but characterized by a series of opaque white superficial stripes and blotches taking the form in typical specimens of (a) a median dorsal line which bifurcates anteriorly, passing on to the oral tentacles, (b) a similar line down each flank, uniting on the metapodium with the median line, (c) a line down the back of each rhinophore from a white tip, (d) a line on the anterior face of each ceras, and (e) a broad band on each ceratal tip. The cerata are arranged in up to 5 latero-dorsal clusters (indistinctly demarcated towards the rear), and each has a digestive lobule of a brilliant red colour. Occasional specimens lack some of the white markings and can be difficult to identify. The shape of the oral tentacles and the size and colour of the cerata (Fig. 77) are then the only reliable external features. It feeds upon gymnoblastic hydroids (chiefly *Tubularia* and *Coryne*) but may take also some calyptoblasts (*Hydrallmania* and *Sertularia*).

This aeolid has been recorded down to considerable depths, to 80 m off British coasts, and to an astonishing 400 m elsewhere. British records are frequent and widely scattered, from Devon, Cornwall and Pembrokeshire to the Isle of Man, Norfolk, and St. Abb's Head in eastern Scotland. Outside Britain little is known of its habitats but it occurs from southern Norway to the Mediterranean Sea.

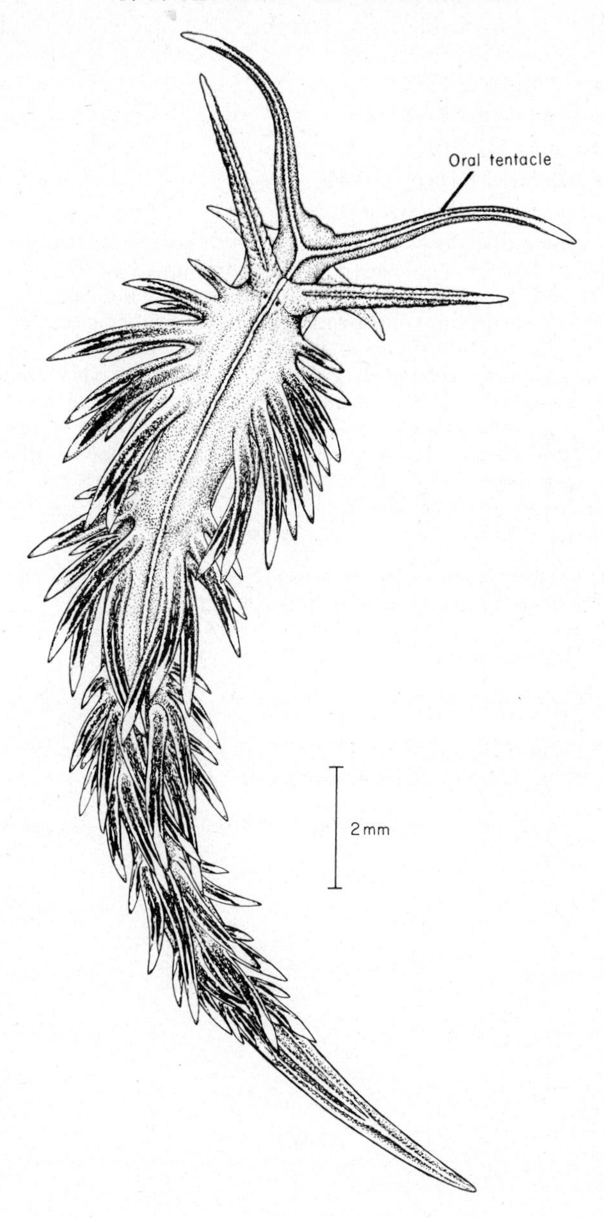

FIG. 77. *Coryphella lineata* dorsal view.

Coryphella verrucosa (M. Sars, 1829) (Figs. 78A and B, 79A and B)

Eolidia verrucosa M. Sars, 1829
Eolis rufibranchialis Johnston, 1832
Eolis pellucida Alder and Hancock, 1843
Eolis gracilis Alder and Hancock, 1844
Eolis smaragdina Alder and Hancock, 1851

This species is somewhat variable and at least 3 well known varieties exist which will be described later. *Coryphella verrucosa* grows up to 62 mm in extended length and the body is transparent white without any longitudinal white streaks, although conspicuous opaque white pigment is present on the tips of the tentacles and the cerata. The cerata form up to 7 or 8 latero-dorsal clusters on each side and each ceras contains a red (rarely green) digestive lobule. This species feeds upon both gymnoblastic (*Tubularia* and *Garveia*) and calyptoblastic hydroids (e.g. *Hydrallmania*) off northern coasts of Britain, especially the North Sea and the Irish Sea, but there are a few records from the English Channel. Elsewhere it has been recorded from Greenland and Iceland to the Baltic Sea and Helgoland to the French Biscay coast, to 300 m.

Three varieties of this species occur in British waters.

Variety 1: *Coryphella verrucosa verrucosa* (Fig. 78A)
Eolis pellucida Alder and Hancock, 1843
Eolis gracilis Alder and Hancock, 1844
Digestive gland lobes pale red, smooth.

Variety 2: *Coryphella verrucosa rufibranchialis* (Johnston, 1832) (Fig. 79A and B)
Digestive gland lobes brilliant brick-red, pustulose, cerata more elongated (especially in juveniles).

Variety 3: *Coryphella verrucosa smaragdina* (Alder and Hancock, 1851) (Fig. 78B)
Digestive gland lobes bright green.

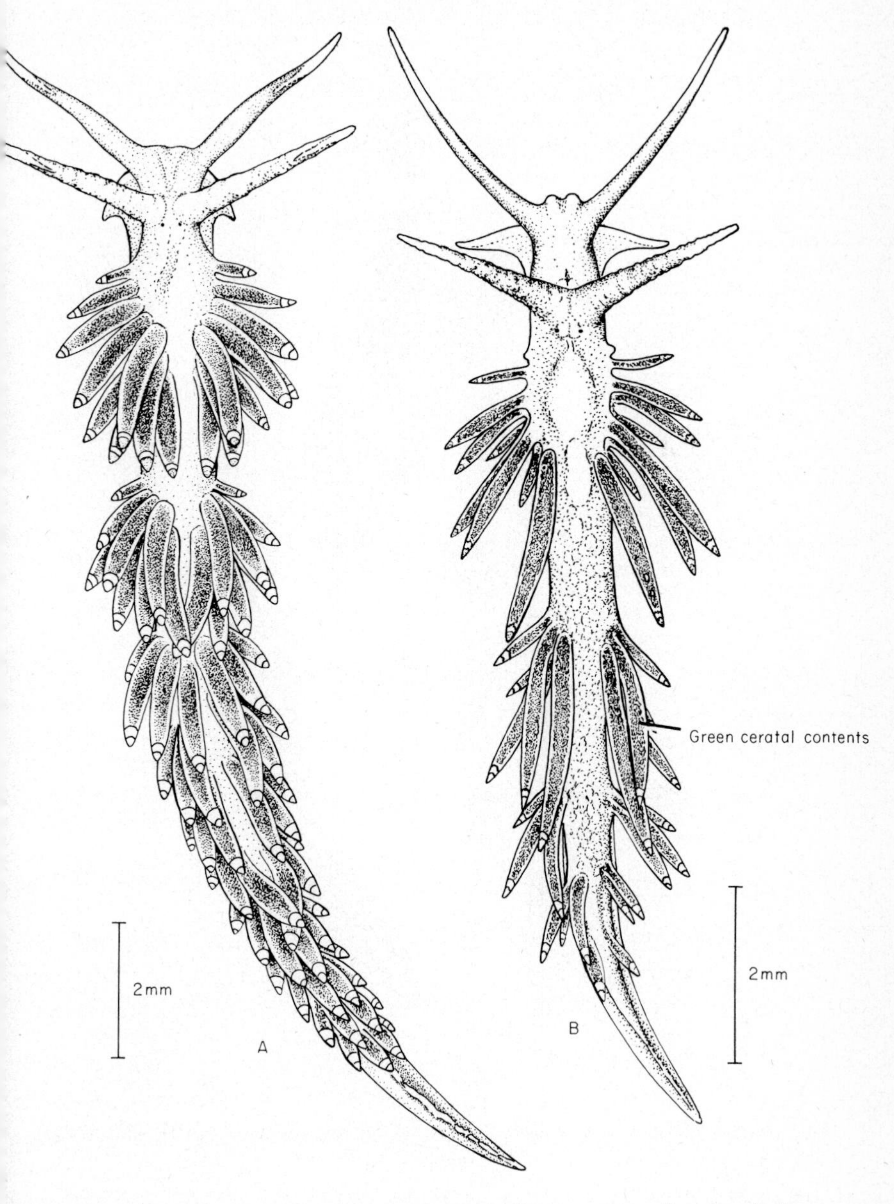

FIG. 78. *Coryphella verrucosa:* A, var. *verrucosa;* B, var. *smaragdina* dorsal views.

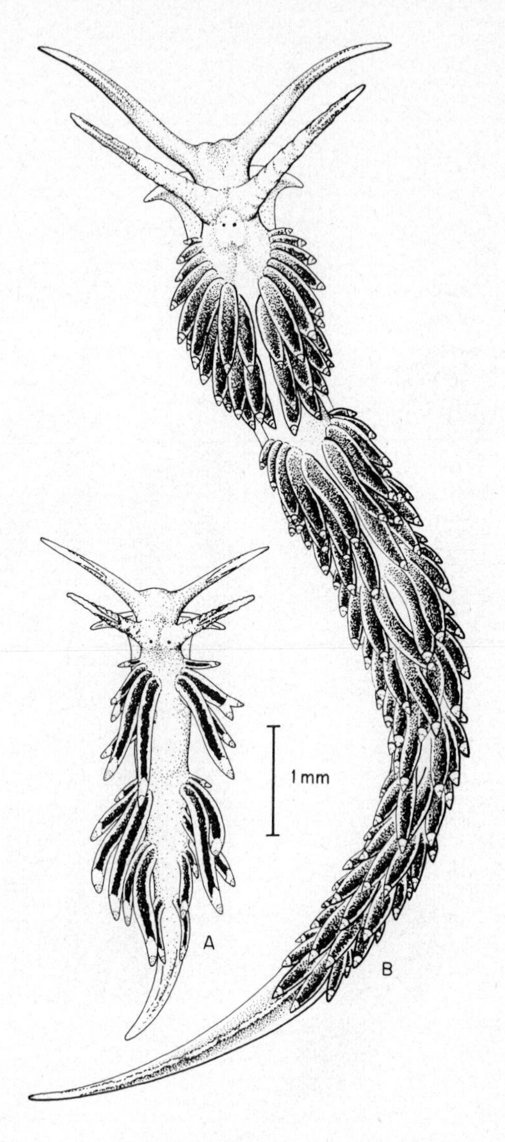

FIG. 79. *Coryphella verrucosa rufibranchialis:* A, juvenile dorsal view; B, adult dorsal view.

Family FACELINIDAE

Genus FACELINA Alder and Hancock, 1855

Facelina auriculata (Müller, 1776) (Figs 80 and 81)

Doris auriculata Müller, 1776
Doris longicornis Montagu, 1808
Eolida coronata Forbes and Goodsir, 1839
Eolis drummondi Thompson, 1844
Eolis curta Alder and Hancock, 1843
Eolis elegans Alder and Hancock, 1845

This slender and beautiful aeolid reaches up to 50 mm in length, but usually about 30 mm. The body is white with a pinkish tinge. The numerous cerata are arranged in up to 8 clusters on either side of the back. They each contain a red, brown, or, rarely, greenish, lobe of the digestive gland; more superficially they exhibit a white subterminal band. The rhinophores each bear up to 30 conspicuous lamellae. The oral tentacles are long and slender (longer than the rhinophores) and conspicuous propodial tentacles are also present.

This energetic, agile and aggressive aeolid is found commonly all around the British Isles, feeding upon *Tubularia* and *Clava* and upon a wide variety of calyptoblastic hydroids. A distinguished naturalist of the nineteenth century, J. G. Dalyell, fed *Facelina* in his laboratory on pieces of mussel and periwinkle and many authors have reported this nudibranch to consume other specimens in aquaria. Outside Britain *F. auriculata* has been recorded from Norway to the Mediterranean Sea, to 120 m.

Facelina auriculata occurs in two morphological varieties, although most authorities consider that sufficient intermediates have been described for it to be unwise to maintain them as separate species. But it is still advisable to record the variety when making biological observations.

Variety 1: *Facelina auriculata coronata* (Forbes and Goodsir, 1839) (Fig. 80)

The body shape is slender, the digestive gland lobules are pinkish and the ceratal epidermis has an iridescent blue appearance.

Variety 2: *Facelina auriculata curta* (Alder and Hancock, 1843) (Fig. 81)

The body shape is broader, the digestive gland lobules are brownish or greenish and the ceratal epidermis has no iridescence.

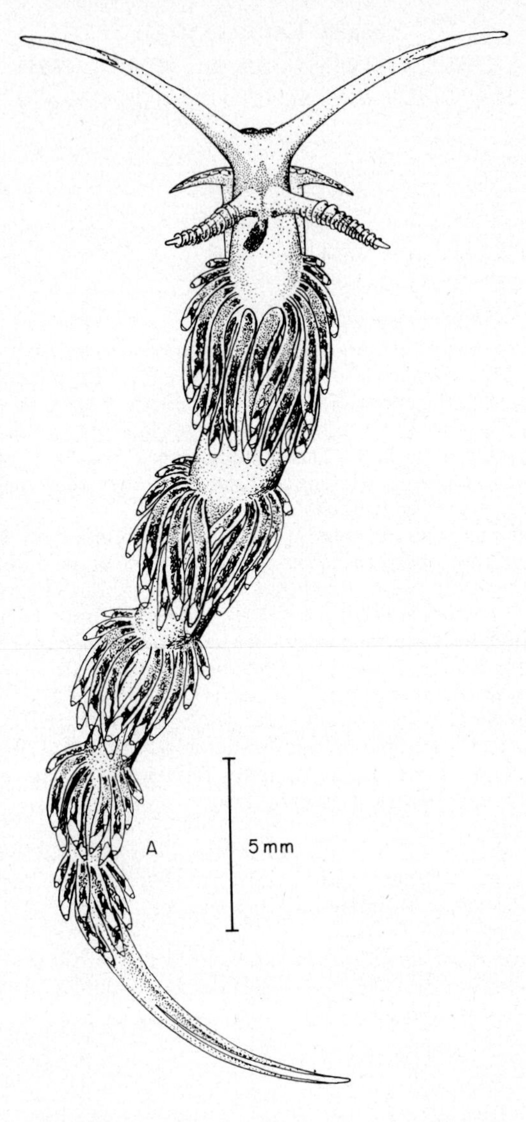

FIG. 80. *Facelina auriculata coronata* dorsal view.

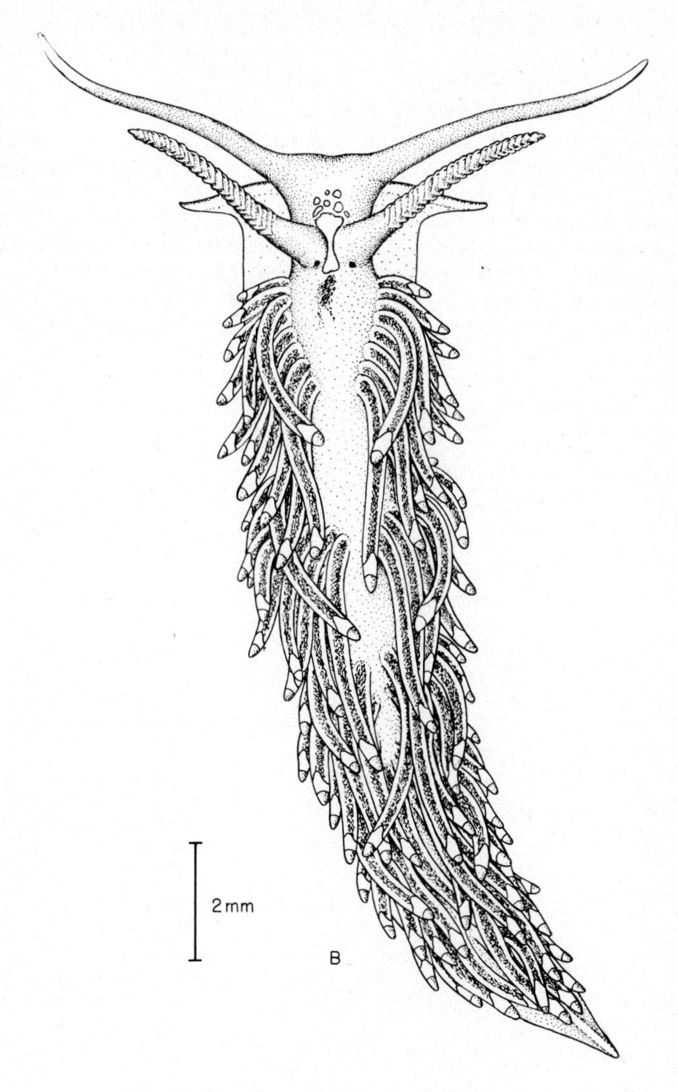

FIG. 81. *Facelina auriculata curta* dorsal view.

Facelina annulicornis (Chamisso and Eysenhart, 1821) (Fig. 82)

Eolidia annulicornis Chamisso and Eysenhart, 1821

Eolis punctata Alder and Hancock, 1845

This species may reach 43 mm in extended length and is even more slender and delicate in appearance than the other British species of *Facelina*. But this disguises an aggressive and restless disposition. The body is brownish pink in colour, with opaque white spots which are sprinkled also over the cerata. Nearly all the brownish cerata exhibit a white distal tip and the oral tentacles are also generously tipped with white. Conspicuous recurved antero-lateral propodial tentacles are present. An obvious feature, peculiar to this species, is that the rhinophores bear unusually fine, even, obliquely sloping lamellae (unlike the "dinner-plates" shown by *Facelina auriculata*).

British records are few and come from Devon and Pembrokeshire. Elsewhere it has been recorded only from the Atlantic and Mediterranean coasts of France.

FIG. 82. *Facelina annulicornis* dorsal view.

Family FAVORINIDAE

Genus FAVORINUS Gray, 1850

Favorinus blianus Lemche and Thompson, 1974 (Fig. 83)

The extended length may reach 30 mm, pale straw-yellow in colour but frequently with black pigment on the rear face of each rhinophore. Opaque white pigment streaks the dorsal sides of the oral tentacles, and of the propodial tentacles and the metapodium, as well as covering the distal tips of the cerata. The digestive gland lobes within the cerata are pale yellow-brown in colour. The numerous cerata are arranged in horse-shoe clusters but this arrangement may be hard to discern; the cerata tend to sweep inwards over the dorsum. Each ceras is flattened in section. The two rhinophores are placed close together and are shorter and stouter than the digitiform oral tentacles. The rhinophores widen a little from the base up to the first of the 3 ring-shaped swellings characteristic of *F. blianus*. Recurved antero-lateral propodial tentacles are present.

This species is known to consume coelenterates such as *Tubularia*. The spawn of other nudibranchs (e.g. *Doto coronata*) is also eaten and fully formed embryos of the prey may be seen within the digestive gland lobes of *F. blianus*. More ecological information is urgently needed.

F. blianus has been recorded from Pembrokeshire, Cornwall, N. Devon, Lundy and western Ireland (Killary Bay), down to 35 m, as well as from Norway and Denmark (Kattegat). It probably also occurs on the Atlantic coast of France.

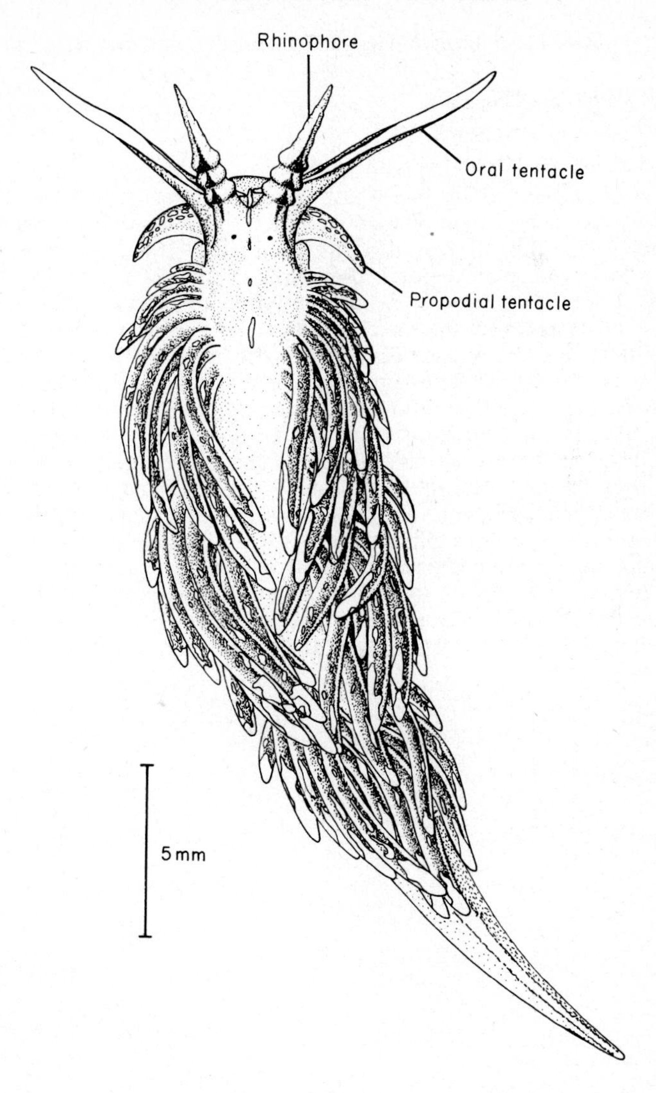

FIG. 83. *Favorinus blianus* dorsal view.

Favorinus branchialis (Rathke, 1806) (Fig. 84A and B)

Doris branchialis Rathke, 1806
Eolis alba Alder and Hancock, 1844
Favorinus albidus Iredale and O'Donoghue, 1923

This is an unusually variable but always slender and fragile species which may reach 25 mm in overall length. The colour varies from pure white to pale brown, sometimes with yellow or rosy digestive diverticula. The most conspicuous feature is the presence on the dark brown rhinophores of a pale tip which is dilated as shown in Fig. 84A. Sometimes each rhinophore bears only one bulb, whereas in other specimens there may be two. One individual has been described in which the rhinophore of one side possessed two bulbs while that of the other side had none. Recurved antero-lateral propodial tentacles are present.

Like *F. blianus*, this species is known to eat both coelenterates (calyptoblastic hydroids and, in the laboratory, small sea anemones) and the eggs of other opisthobranch molluscs (such as *Aplysia*, *Tylodina*, *Polycera* and *Archidoris*). The colour of the prey soon affects that of the digestive gland of *Favorinus*. This dietary factor, combined with the variable distribution of superficial white and brown pigment, makes this a difficult species to describe simply. In addition both bodily coloration and rhinophoral bulbs may be lacking in juveniles.

There are not many British records but they are widely scattered around our coasts. Further distribution from Murmansk and Norway to the Mediterranean Sea, to 20 m. Records from Brazil, West Indies and Florida need confirmation.

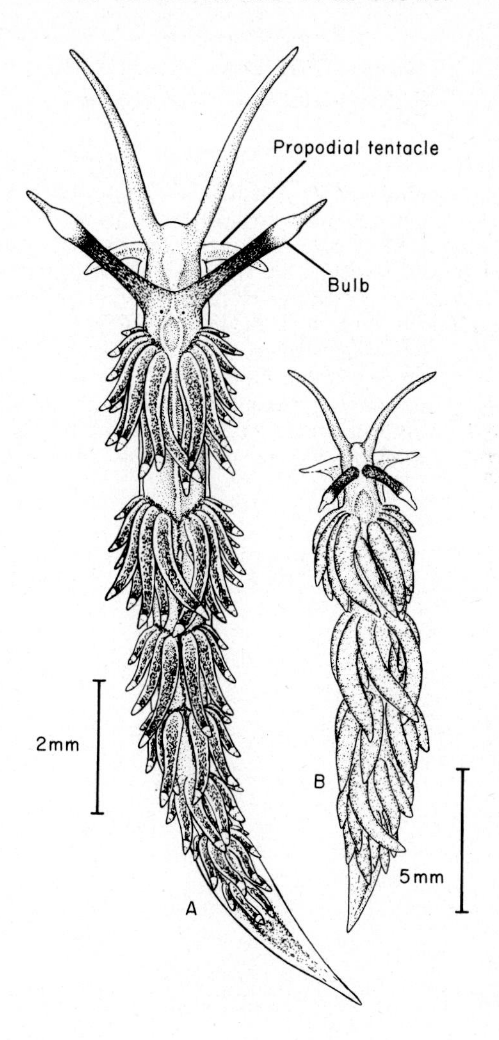

FIG. 84. *Favorinus branchialis:* A, dark variety dorsal view; B, pale variety dorsal view

Family AEOLIDIIDAE

Genus AEOLIDIA Cuvier, 1798

Aeolidia papillosa (L., 1761) (Fig. 85)

Limax papillosus L., 1761

This is the largest aeolid of the N. Atlantic and reaches 120 mm in length. The body is variable in colour, from white to dark purple-brown; the central bare area of the dorsum usually bears pale markings often forming a distinctive white crescent over the head. The cerata are elongated and flattened, with white tips. They are arranged in 25 or more transverse rows. The rhinophore bases are usually darker than the rest of the body while the tips are pale.

This world-wide species has been recorded from shores and shallow waters all around the British Isles where it feeds on sea anemones (*Sagartia*, *Actinia*, *Metridium*, *Tealia* and *Anemonia*), taking large bites from the stipe of the prey. Elsewhere it has been recorded from the White Sea to the French Biscay coast as well as from New England, Vancouver and Japan to great depths (800 m).

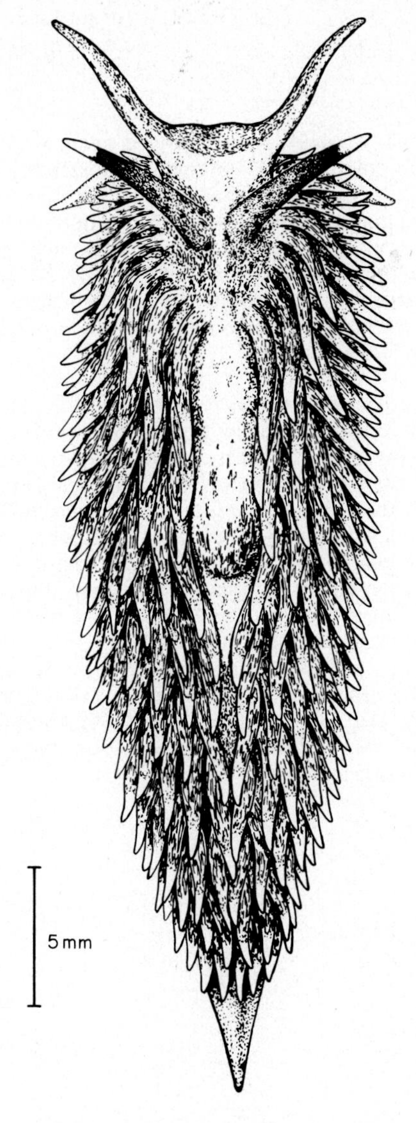

FIG. 85. *Aeolidia papillosa* dorsal view.

Genus AEOLIDIELLA Bergh, 1867

Aeolidiella alderi (Cocks, 1852) (Fig. 86)

Eolis alderi Cocks, 1852

This active species may reach 37 mm in extended length, and is flattened in shape, the body intrinsically colourless but the coloured alimentary canal may be visible through the body wall. The diet certainly governs such coloration. The digestive lobules may be pale pink to dark brown. The cerata are set in up to 14 closely-packed transverse rows on either side of the back; the first 1 or 2 rows are characteristically colourless and form a "ruff" behind the rhinophores. The oral and rhinophoral tentacles are smooth (i.e. non-lamellate, although transient surface wrinkling can occasionally be seen) and the propodium is produced to form curved antero-lateral tentacles.

It is certain that 3 species of *Aeolidiella* occur in British waters but they have in the past been confused and the individual geographical distributions are unclear. All three also occur on the Biscay coast of France where Tardy (1969) has studied them in detail and has patiently elucidated the morphological and behavioural distinctions between them. The reader is advised to consult Dr. Tardy's paper for fuller information but should be warned that successful identification of a single specimen cannot be guaranteed. All three species feed upon sea-anemones and all three have broadly similar external morphology. But the white "ruff" of *A. alderi* is less evident in the other two species. One of these, *A. glauca* (Alder and Hancock, 1845) shows instead scattered white pigment on the back, especially on the rhinophore tips and the other tentacles. The other species, *A. sanguinea* (Norman, 1877) cannot be certainly identified at present without dissection.

There is an urgent need for further records and observations of these 3 species of *Aeolidiella* so that their British ranges can be accurately mapped. The present situation is far from satisfactory.

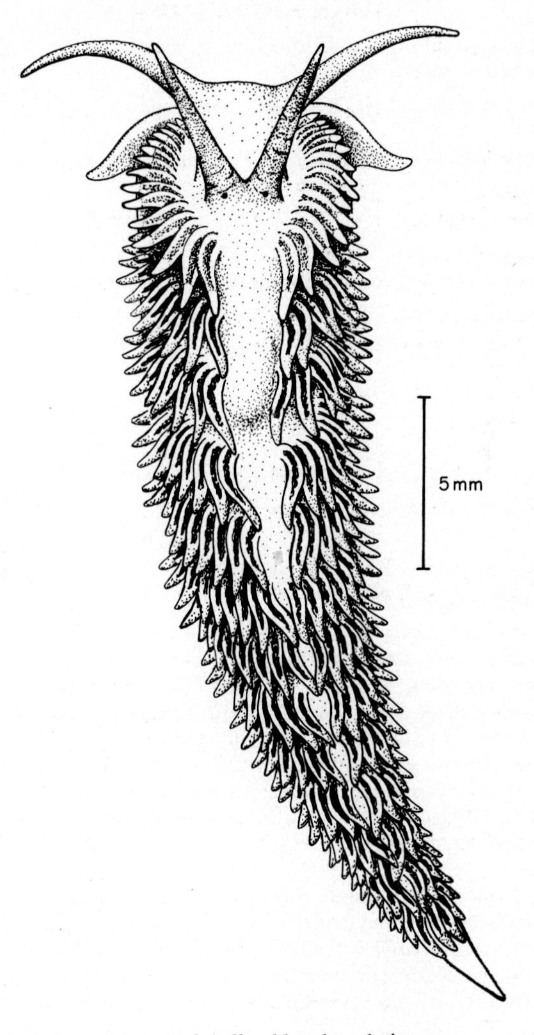

FIG. 86. *Aeolidiella alderi* dorsal view.

Family EUBRANCHIDAE

1. Cerata each exhibiting a golden yellow subterminal ring **2**
   Cerata without such a ring **3**
2. Rhinophores and oral tentacles with bold yellow-orange pigment *Eubranchus farrani* (p. 166)
   Tentacles without such yellow-orange pigment . *Eubranchus tricolor* (p. 164)
3. Cerata inflated **4**
   Cerata more slender . . . . *Eubranchus vittatus* (p. 168)
4. Pigmentation greenish brown **5**
   Pigmentation red-brown with no trace of green . *Eubranchus pallidus* (p. 168)
5. Cerata smooth, greatly inflated and club-shaped. *Eubranchus exiguus* (p. 170)
   Cerata knobbly, less obviously inflated . . *Eubranchus cingulatus* (p. 166)

Genus EUBRANCHUS Forbes, 1838

Eubranchus tricolor Forbes, 1838 (Fig. 87)

Eolis violacea Alder and Hancock, 1844

Galvina viridula Bergh, 1873

The body length may reach 45 mm, pale yellow or greyish white in colour. The inflated cerata are not arranged strictly in rows but are closely packed together (early authorities often wrote of its superficial resemblance to a hedgehog), especially in the adults where the cerata are crowded to such an extent that they are flattened fore and aft. They are characteristically translucent white or cream with a golden yellow subterminal ring surrounded by opaque white pigment. The digestive diverticula are reddish brown, often showing a tinge of violet towards the tip. Propodial tentacles are absent and the rhinophores are smooth. It feeds upon both gymnoblastic (*Tubularia*) and calyptoblastic (e.g. *Hydrallmania*) hydroids.

Because of uncertainty about some of the early records it is impossible at present to give accurately the British distribution. Further information is urgently needed. *Eubranchus tricolor* was first described from the Irish Sea, to 80 m, and it certainly occurs off Northumberland and in the English Channel. Elsewhere it is known from N. America, Greenland and the White Sea to Brittany.

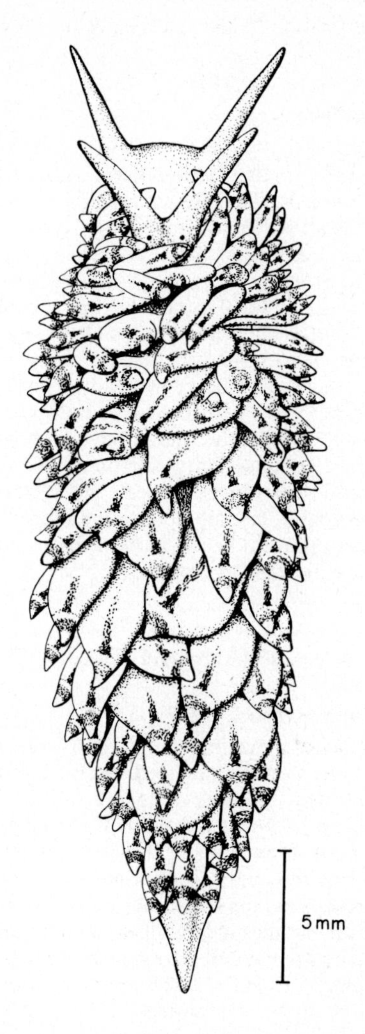

FIG. 87. *Eubranchus tricolor* dorsal view.

Eubranchus farrani (Alder and Hancock, 1844) (Fig. 88)

Eolis farrani Alder and Hancock, 1844
Amphorina alberti Quatrefages, 1844

This species has in the past often been confused with *Eubranchus tricolor* but *E. farrani* is more slender, smaller, with fewer cerata or less compressed shape and different colour patterns. The present species does not exceed 20 mm in extended length and has an elongated appearance (Fig. 88).

The typical *E. farrani* coloration is a translucent white body with orange-tipped rhinophores and oral tentacles, with scattered orange spots and blotches on the dorsum, and white inflated cerata having a conspicuous sub-terminal orange or yellow ring. But occasional variants show individual exaggeration of certain of the markings. The entire body, cerata and all, may be of a beautiful golden hue. In others there may be brown blotches on the body and in one specimen (from Pembrokeshire) nearly the whole body was dark chocolate brown in colour. Juveniles may, on the other hand, have no pigment at all.

Like the previous species, uncertainty surrounds many faunistic records for *E. farrani*, and more information is needed. But it appears that this species occurs off most British coasts, feeding upon calyptoblastic hydroids (e.g. *Obelia geniculata*) attached to sublittoral kelps, to 30 m. Elsewhere it has been reported from Norway to the Mediterranean Sea.

Eubranchus cingulatus (Alder and Hancock, 1847) (Fig. 89)

Eolis cingulata Alder and Hancock, 1847

This delicate species reaches a maximal length of 29 mm and is regarded as rare possibly due to its being a well-camouflaged species and therefore easily missed. The body is grey-white with blotches of superficial olive green or brown. The cerata are arranged in up to 9 transverse rows and stand up in fan-like array. They bear olive green or brown pigment which sometimes is aggregated to form 3 bands. Each ceras may vary from minute to minute in the degree of its elongation. Sometimes a ceras may contract and then its shape appears to be distinctly knobbly; at other times it can appear smooth and digitiform. The oral and rhinophoral tentacles are banded transversely (Fig. 89).

This species has been recorded from scattered localities all around the British Isles, to 40 m, feeding upon calyptoblastic hydroids (e.g. *Kirchenpaueria*). Elsewhere it is known to occur from Norway to the French Biscay coast, to 40 m.

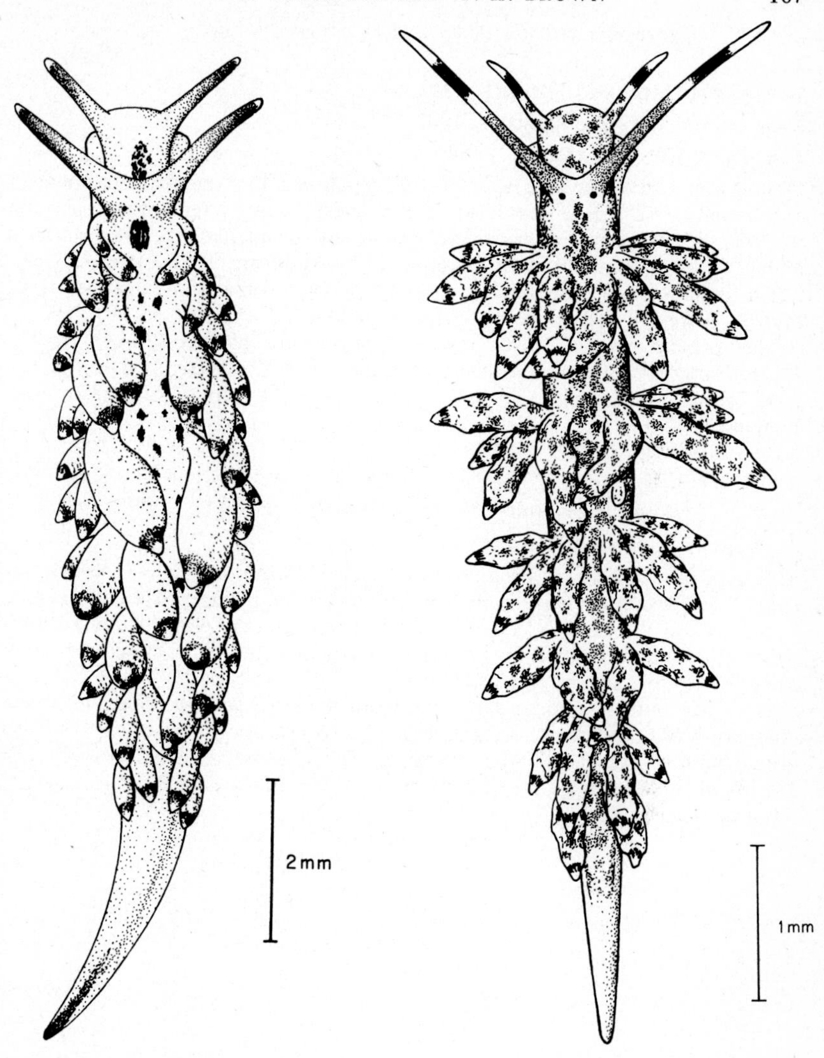

FIG. 88. *Eubranchus farrani* dorsal view.

FIG. 89. *Eubranchus cingulatus* dorsal view.

Eubranchus pallidus (Alder and Hancock, 1842) (Fig. 90A)

Eolis pallida Alder and Hancock, 1842
Eolis minuta Alder and Hancock, 1842
Eolis picta Alder and Hancock, 1847

Despite its name this species is not especially pale in colour, but well camouflaged and therefore may easily be missed. It may reach 23 mm in length, with a white body mottled dorsally with brown and orange and the inflated cerata are similarly pigmented (occasionally paler juvenile individuals are encountered). These latero-dorsal ceratal processes are more obviously arranged in transverse rows than in *E. farrani* and, more especially, *E. tricolor*.

The present species feeds upon calyptoblastic hydroids (e.g. *Obelia, Hydrallmania*) in offshore localities all around the British Isles, to 60 m. Further distribution from Iceland and the White Sea down to Brittany; and from New England.

Eubranchus vittatus (Alder and Hancock, 1842) (Fig. 90B)

Eolis vittata Alder and Hancock, 1842

This very rare species has been reputed to reach 19 mm in extended length but has not been examined alive in British waters in living memory. The body is pale buff in colour with rust-coloured blotches. The cerata are elongated and digitiform, buff-coloured with 2 or 3 rust-coloured bands and pale yellow tips. They are set in up to 7 transverse rows.

Only two reliable records exist, one from the Irish Sea, the other from the coasts of Northumberland, on hydroids. This species and *E. cingulatus* (Fig. 89) are similar in all except colour pattern and ceratal shape but until further specimens are found, it will not be possible to decide if *E. vittatus* and *E. cingulatus* are separate species.

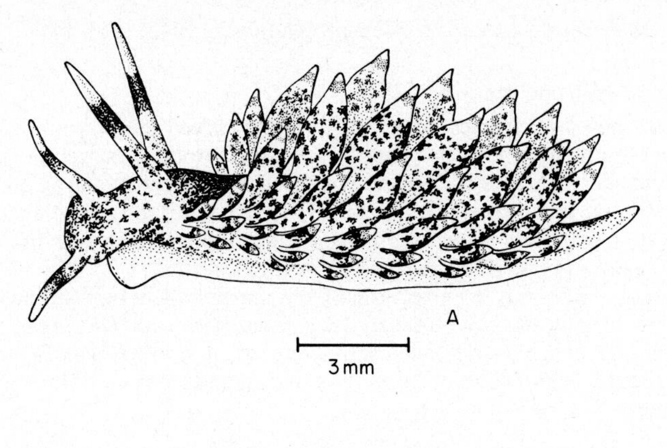

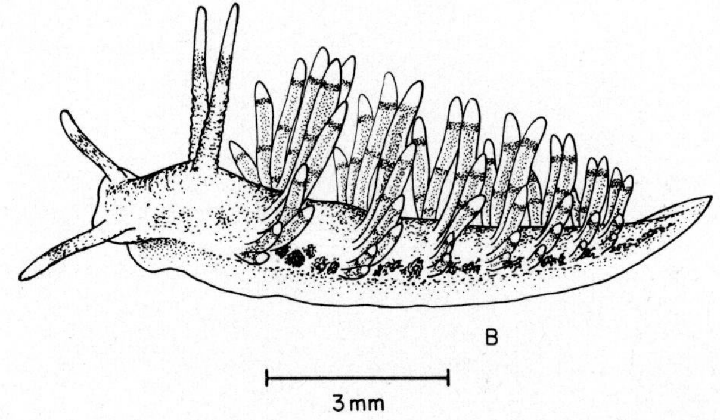

FIG. 90. *Eubranchus pallidus:* A, lateral view; *Eubranchus vittatus:* B, lateral view (both after Alder and Hancock (1845)).

Eubranchus exiguus (Alder and Hancock, 1948) (Fig. 91)

Eolis exigua Alder and Hancock, 1848

This delicate but drab species does not normally exceed 10 mm in adult length and is often found spawning at a length of 5 mm. The body is grey or yellowish white with spots or patches of brown. The rather sparse cerata (up to 10 on each side) are inflated and urn-shaped (Fig. 91), but not knobbly like those of *E. cingulatus* (Fig. 90). The brown pigment on the skin of each ceras often forms interrupted bands (as in *E. cingulatus*).

The present species has been recorded from scattered localities all around the British Isles, usually on calyptoblastic (e.g. *Laomedea* and *Obelia*) or gymnoblastic (*Coryne*) hydroids on the lower shore or in shallow waters, to 40 m. Outside Britain it is known to occur in N. America and from the White Sea to the Mediterranean Sea, to 140 m.

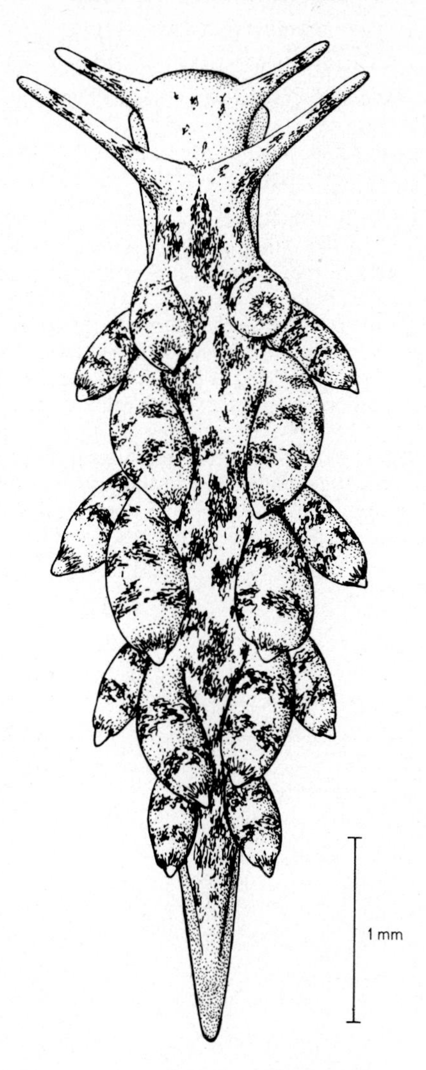

FIG. 91. *Eubranchus exiguus* dorsal view.

Family CUMANOTIDAE

Genus CUMANOTUS Odhner, 1907

Cumanotus beaumonti (Eliot, 1906) (Fig. 92)

Coryphella beaumonti Eliot, 1906
Cumanotus laticeps Odhner, 1907

This species is rare in British waters; it may reach 20 mm in extended length. It is the only aeolid of the Atlantic Ocean which can swim, which it does by repeated violent vertical movements of the cerata (shown in Fig. 92). This appears to be an escape response. The body is reddish with flesh-coloured cerata, yellowish towards the tips. The digestive lobes are reddish, rarely greenish. The cerata are arranged in numerous transverse rows, several of which are in front of the smooth rhinophores. The head is broad and bears a pair of finger-like oral tentacles. The propodium is produced laterally to form a pair of recurved propodial tentacles.

The species is so rare in British waters that none has been found for over half a century. Very few Atlantic records exist of this species, all from the Devon coast or from Sweden. A species of *Cumanotus* has been found on the Pacific coast of north America, however, and some authorities consider that it is identical with *C. beaumonti*.

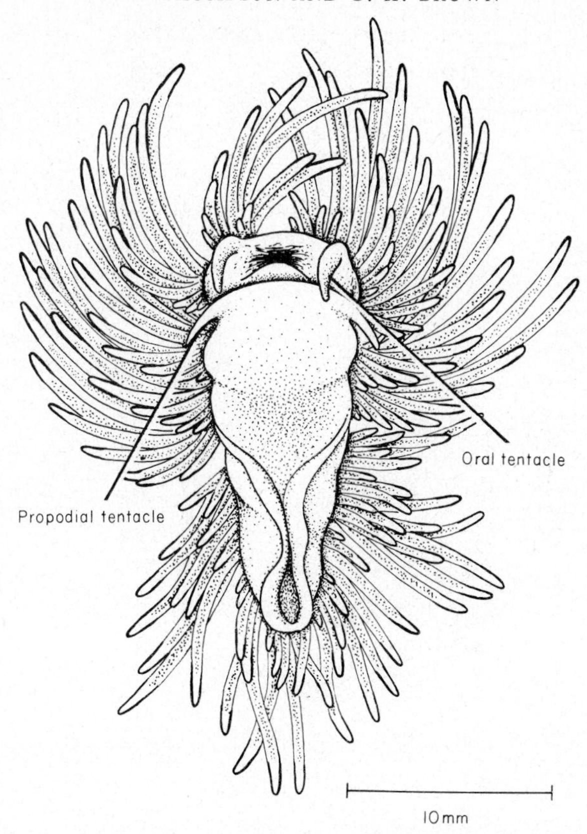

FIG. 92. *Cumanotus beaumonti* ventral view, swimming (drawing taken from original photograph of Puget Sound (U.S.A.) specimen).

Family PSEUDOVERMIDAE

Genus PSEUDOVERMIS Pereyaslawzewa, 1891

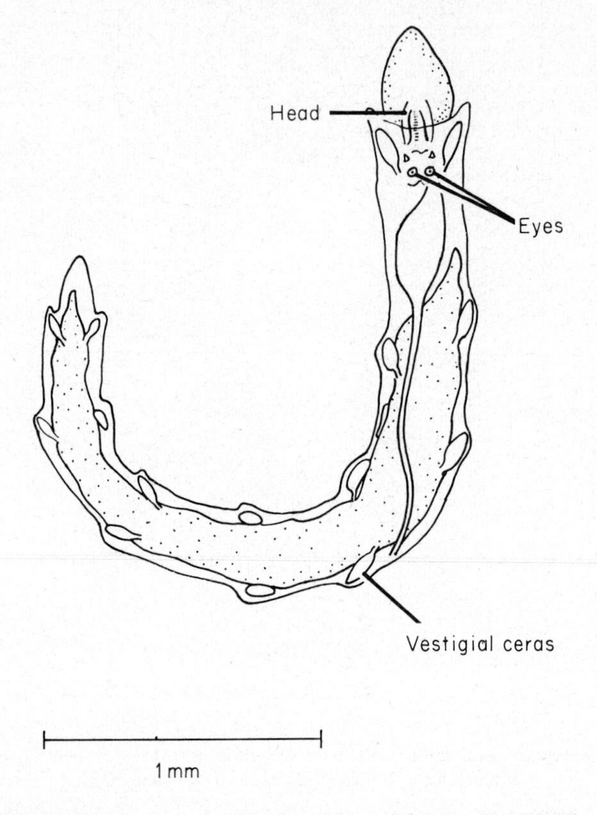

FIG. 93. *Pseudovermis boadeni* dorsal view (after Boaden (1961)).

Pseudovermis boadeni Salwini-Plawen and Sterrer, 1968 (Fig. 93)

This is the sole British representative of a family with numerous European species. Only 3 specimens have ever been found (on the east coast of Anglesey) and the largest of these measured only 3·5 mm in length. They look more like worms than molluscs but the presence of an internal radula (verifiable only with a microscope) and of latero-dorsal ceratal tubercles like those of a typical aeolid, show this to be a nudibranch. It has become adapted to life in the interstices of littoral coarse sand, preying upon slender cnidarians such as *Halammohydra vermiformis*. The only known specimens were found by patient sieving and searching through quantities of fine shell-gravel.

Family CUTHONIDAE

1. Oral tentacles present **2**
Oral tentacles absent . . . . *Embletonia pulchra* (p. 179)

2. Head rounded, shield-like bearing short, posteriorly direct oral tentacles *Tenellia pallida* (p. 180)
Head not shield-like; oral tentacles elongated and laterally or anteriorly directed **3**

3. Cerata with strong blue areas . . . . . . *Trinchesia caerulea* (p. 186)
Cerata without blue areas **4**

4. Cerata with red or pink contents **5**
Cerata not red or pink **7**

5. Cerata extend forwards anterior to the rhinophoral bases **6**
Cerata do not reach so far as the rhinophoral bases *Catriona aurantia* (p. 182)

6. Almost invariably found on *Hydractinia echinata* on hermit crab shells *Precuthona peachi* (p. 176)
Habits unknown but not as above . . . . . . *Cuthona nana* (p. 178)

7. Cerata few (less than 10) . . . . *Tergipes tergipes* (p. 184)
Cerata abundant (more than 20) **8**

8. Rhinophores each with distinct brownish pigment band *Trinchesia amoena* (p. 188)
Rhinophores more evenly coloured **9**

9. Ceratal contents green . . . . *Trinchesia viridis* (p. 190)
Ceratal contents grey, yellow or brown **10**

10. Dorsal surface of head with conspicuous pinkish linear markings *Trinchesia foliata* (p. 192)
No such markings . . . . *Trinchesia concinna* (p. 190)

Genus PRECUTHONA Odhner, 1929

Precuthona peachi (Alder and Hancock, 1848) (Fig. 94)

Eolis peachii Alder and Hancock, 1848

The body of this interesting species ranges up to 22 mm in length, usually not more than 12 mm, pale cream or yellow in colour but with a red blush over the posterior half of the dorsum. The cerata each possess a central brown digestive diverticulum lobule and a white tip. The cerata are numerous, arranged in about 20 latero-dorsal rows on each side of the back. The oral and rhinophoral tentacles are smooth; the propodium bears no tentacular processes.

This species has very precise food requirements and is almost invariably found feeding upon the hydroid *Hydractinia echinata* growing epizoitically on molluscan shells inhabited by hermit crabs (especially *Pagurus bernhardus*). In this situation it is well camouflaged and can only be detected by bringing into the laboratory colonies of *Hydractinia* and examining these under a binocular microscope. The spawn of this nudibranch is also coloured so as to elude detection.

Precuthona peachi has been recorded from Devon and Cornwall, and from the Irish and North Seas to 60 m. Elsewhere it is known only from Sweden.

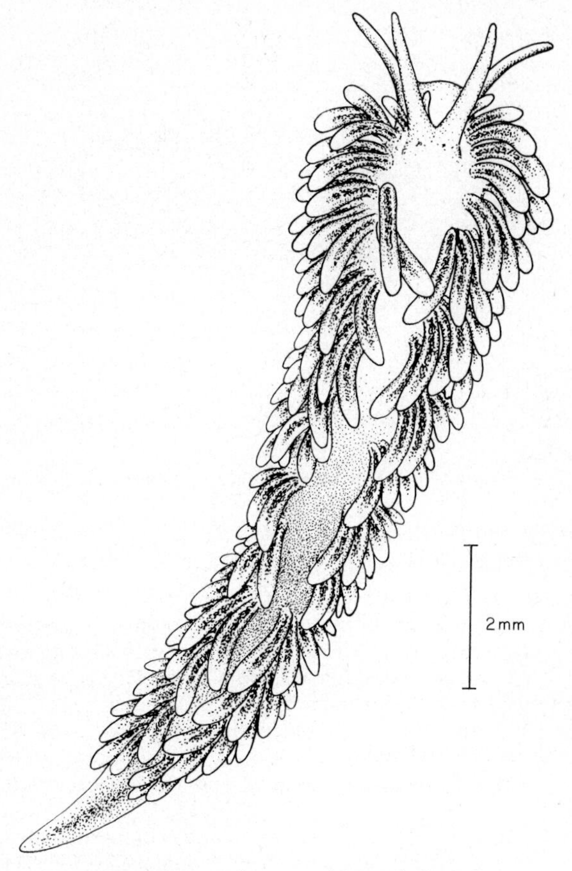

FIG. 94. *Precuthona peachi* dorsal view.

Genus CUTHONA Alder and Hancock, 1855

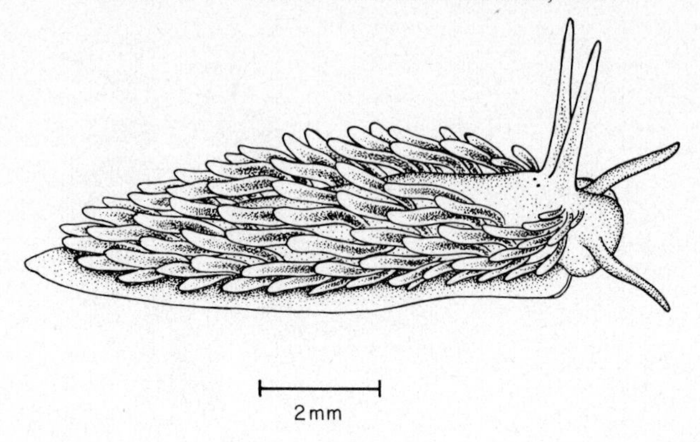

FIG. 95. *Cuthona nana* lateral-dorsal view (after Alder and Hancock (1845–55)).

Cuthona nana (Alder and Hancock, 1842) (Fig. 95)

Eolis nana Alder and Hancock, 1842

Eolis pustulata Alder and Hancock, 1854

This rare species may reach a maximal length of 12 mm. The body colour is white or pale yellow while the cerata contain brown, orange or rose-red digestive lobules. Each ceras may bear white superficial specks (this denotes the variety which the original authors considered to be a separate species *pustulata*). The cerata are arranged in up to 10 transverse rows on either side of the dorsum; they are finger-like and have white obtuse tips. The foot is broad and semicircular in outline. The oral tentacles arise from the mid-dorsal area of the head (not from its lateral extremities) and they and the smooth rhinophores are white-tipped.

Cuthona nana is exceedingly rare and many features of its anatomy are obscure. Over half the existing records have come from the intertidal zone (Northumberland, Cheshire) and this gives hope that renewed searches may be successful. Outside Britain this species is known only from Scandinavia, although there is one doubtful record from the Pacific coast of N. America.

Genus EMBLETONIA Alder and Hancock, 1851

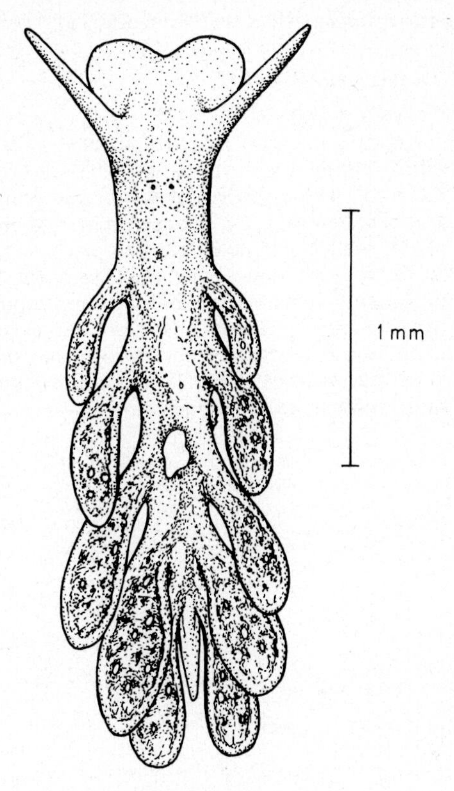

FIG. 96. *Embletonia pulchra* dorsal view.

Embletonia pulchra (Alder and Hancock, 1951) (Fig. 96)

Pterochilus pulcher Alder and Hancock, 1844
Embletonia faurei Labbé, 1923

The body may reach 6 mm in length, pale pink in colour with dorsal white blotches. The digestive lobules vary from pale yellow to dark red-brown, usually pinkish, and the club-shaped cerata form a single series (up to 6) on each side of the body. The frontal margin of the head is characteristically bilobed but lacks oral tentacles. The smooth rhinophores are set wide apart (Fig. 96).

This delicate aeolid occurs on intertidal and shallow sublittoral hydroids (e.g. *Hydrallmania*, *Nemertesia*) off the Isle of Man and various parts of Scotland, but there are also rare records for the English Channel. Elsewhere it has been reported from Norway to Brittany, and there is a record from the Bay of Naples.

Genus TENELLIA Costa, 1866

Tenellia pallida (Alder and Hancock, 1842) (Fig. 97)

Eolis pallida Alder and Hancock, 1842

Embletonia pallida Alder and Hancock, 1854

This drab aeolid attains a maximal body-length of 7 mm, yellowish white in ground colour, with a varying number of patterns of dark brown or black specks on the skin of the back. The digestive lobes are pale orange-pink, each lobe lying inside a ceras. These cerata are arranged in 4–5 latero-dorsal rows. The frontal margin of the head is characteristically domed and bears conspicuous lateral oral tentacles. It feeds upon hydroids (*Laomedea* and *Cordylophora*).

Tenellia pallida is a euryhaline species and has been reported from brackish areas (such as the Dee estuary and the Bristol Channel) around the British Isles. Further distribution is recorded from Norway and the Baltic Sea to the Atlantic and Mediterranean coasts of France, in shallow water down to a salinity of 3·1‰. A report from Japan needs confirmation.

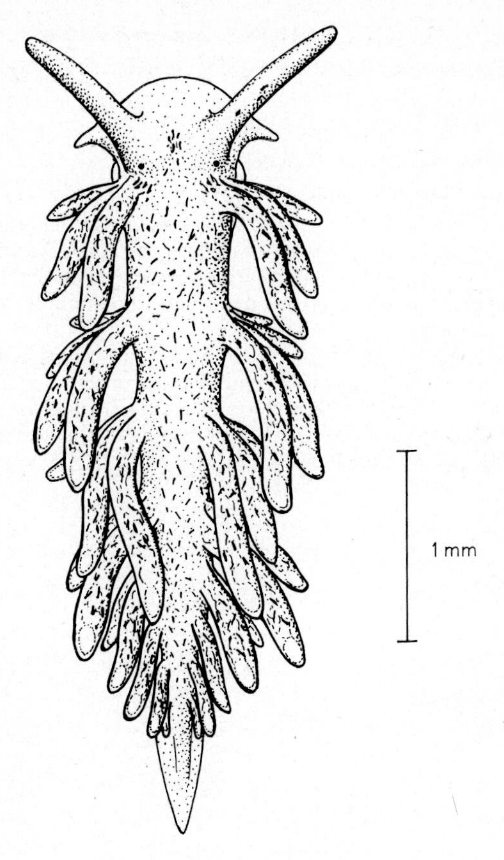

FIG. 97. *Tenellia pallida* dorsal view.

Genus CATRIONA Winkworth, 1941

Catriona aurantia (Alder and Hancock, 1842) (Fig. 98)

Eolis aurantia Alder and Hancock, 1842
Trinchesia aurantia Alder and Hancock, 1842
Eolis aurantiaca Alder and Hancock, 1851

This aeolid may reach 21 mm in length, pale with rosy or orange rhinophores and reddish orange cerata. The cerata are slender and finger-like, set in up to 11 transverse rows on either side of the dorsum. The rhinophoral and oral tentacles are approximately equal in length. The propodium is smoothly rounded and bears no tentacles.

This species is nearly always found feeding on gymnoblastic hydroids (especially *Tubularia* spp.), on which it is well camouflaged. It may occur from the lower shore down to 30 m, and has been recorded from scattered localities all around the British Isles, as well as from New England, and Norway to the Mediterranean Sea. A recent record from the Puget Sound area of the U.S.A. needs confirmation.

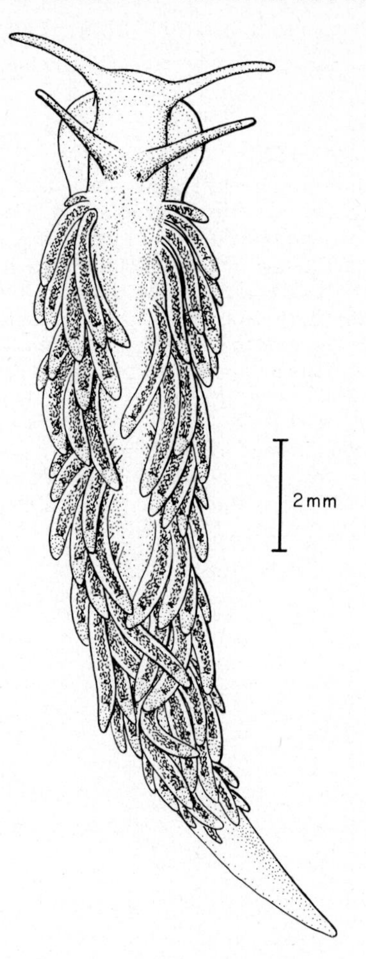

FIG. 98. *Catriona aurantia* dorsal view.

Genus TERGIPES Cuvier, 1805

Tergipes tergipes (Forskål, 1775) (Fig. 99)

Limax tergipes Forskål, 1775
Eolidia despecta Johnston, 1835

This slender and delicate species may reach 8 mm in length but is usually smaller, up to 5 mm. Its appearance is unmistakable and it creeps along in a jerky fashion unlike any other nudibranch. Up to 8 cerata may be present, the first two opposite, the others alternating. The basic colour of the body is translucent white but the undulating digestive gland shows greenish through the skin of the dorsum and the cerata. Each ceras has a white tip and a reddish sub-terminal ring. The rhinophores are more than double the length of the oral tentacles. The skin near the rear of each rhinophore is characteristically reddish and usually joins with a lateral red line down either side of the body.

This species is common all around the British Isles on sublittoral calyptoblastic hydroids (*Obelia*, *Laomedea*), on kelp or on moored rafts or ships, often in company with *Eubranchus exiguus* (Alder and Hancock, 1848). It can tolerate reduced salinity, down to 10‰. Further distribution from Iceland and Norway to the Mediterranean Sea, eastern N. America, Brazil and New Caledonia.

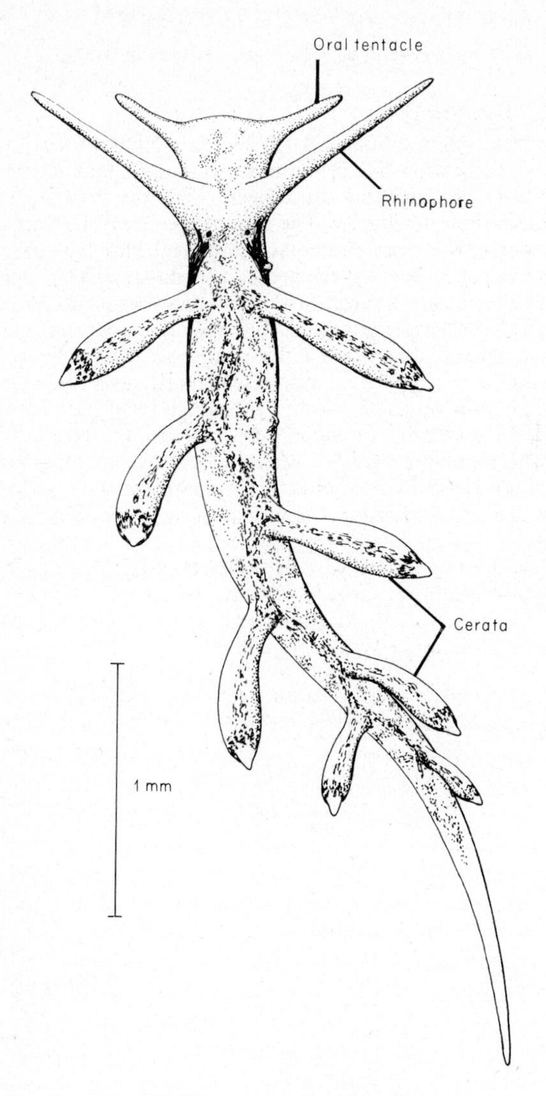

FIG. 99. *Tergipes tergipes* dorsal view

Genus TRINCHESIA Ihering, 1879

Trinchesia caerulea (Montagu, 1804) (Fig. 100A–C)

Doris caerulea Montagu, 1804

This strikingly beautiful nudibranch may reach a length of 18 mm; it is extremely variable and the suggestion has in the past been made that *T. caerulea* and *T. viridis* (Forbes) are one and the same species. For the present, however, it is better to maintain these separately. The skin of the body of *T. caerulea* is white tending to greenish, the rhinophores and oral tentacles are smooth, and the propodium is rounded, lacking tentacles. The most striking and diagnostic features of this species are the cerata. These are set in up to 10 rows and are stout, swollen and brilliantly coloured. The colour derives from the brownish or green-brown digestive diverticula within, but, more obviously, from the strong surface pigments. In the typical condition (excluding juveniles which are paler) there is a brilliant blue ring with a yellow or orange band both above and below it (Fig. 100C). In a commonly encountered variant the cerata and the head tentacles are longer and more slender, while the lower yellow/orange ceratal band is always lacking. These two varieties occur in British and in Mediterranean localities; it may well be necessary to elevate them to the rank of separate and distinct species.

Trinchesia caerulea feeds upon calyptoblastic hydroids (e.g. *Sertularella*, *Hydrallmania*) from the lower shore to considerable depths (70 m around Britain) and has been recorded sporadically from localities all around our coasts. Further distribution from Norway to Naples, to 250 m.

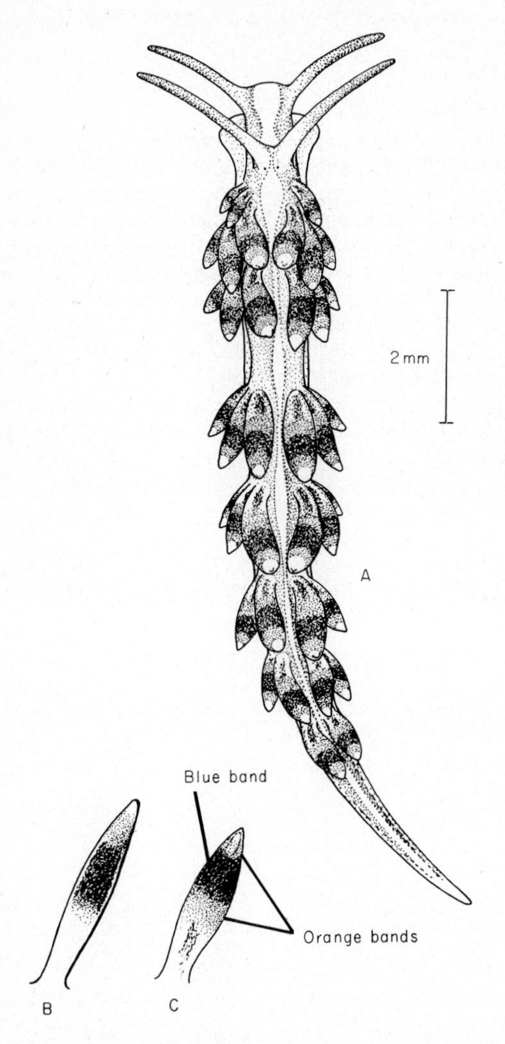

FIG. 100. *Trinchesia caerulea:* A, whole animal dorsal view; B and C, alternative varieties showing pigmentation of cerata.

Trinchesia amoena (Alder and Hancock, 1845) (Fig. 101A and B)

Eolis amoena Alder and Hancock, 1845

Cuthona amoena (Alder and Hancock, 1845)

This delicate aeolid does not exceed 10 mm in extended length, greenish white in body-colour with scattered opaque white spots, sometimes set on tiny papillae in the head region. The digestive diverticula are green-brown in colour, giving the cerata their olive appearance. Characteristically the cerata also exhibit superficial white spots, a white tip, red-brown basal pigment and scattered brown blotches that in some individuals unite to form one or more indistinct rings. The rhinophores are slightly wrinkled and usually have a brown zone near the base; such a brown zone may also be present on each oral tentacle. The propodium is expanded laterally but lacks tentacles. It appears to feed upon calyptoblastic hydroids (especially *Halecium*).

Trinchesia amoena has been recorded from the Bristol Channel, the North Sea, the Irish Sea and the English Channel, to 40 m. Elsewhere there are reports from the Mediterranean Sea.

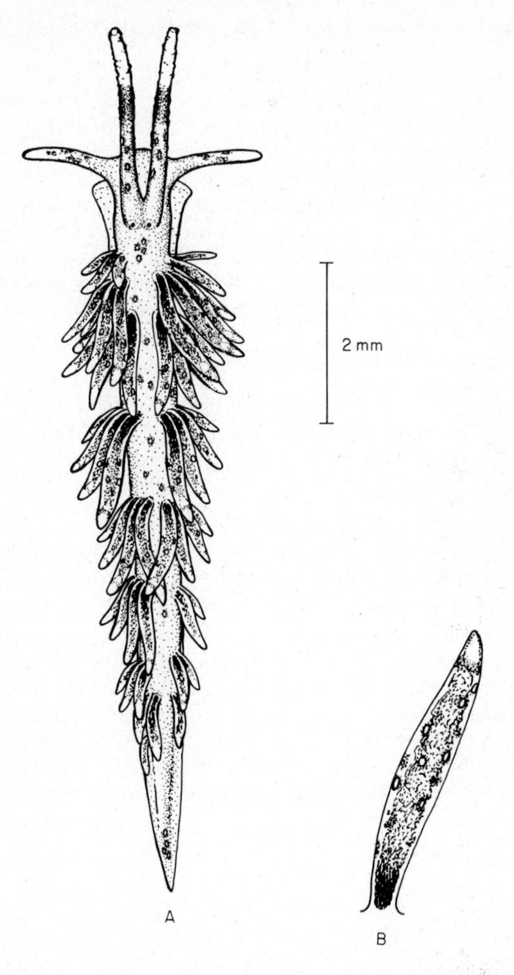

FIG. 101. *Trinchesia amoena:* A, whole animal dorsal view; B, single ceras.

Trinchesia concinna (Alder and Hancock, 1843) (Figs. 102B)

Eolis concinna Alder and Hancock, 1843

Cuthona concinna (Alder and Hancock, 1843)

This rare aeolid may attain a length of 15 mm, cream or yellow in body colour with white-tipped rhinophores. The cerata are numerous, finger-like and arranged in up to 12 rows. The digestive lobules within the cerata are brownish but characteristically have a violet tinge. The tips are white. The oral and rhinophoral tentacles are smooth; the propodium is not produced to form tentacles.

Very few records exist of *T. concinna*, nearly always in association with species of the calyptoblastic hydroid *Sertularia*. It is reported from Northumberland, Lancashire and the Isle of Man, to 25 m. Elsewhere it is known to occur in Netherlands and Norwegian waters. There are unconfirmed records from Vancouver Island, New England and from the White Sea.

Trinchesia viridis (Forbes, 1840) (Fig. 102A)

Montagu viridis Forbes, 1840

Eolis northumbrica Alder and Hancock, 1844

This delicate and interesting species may reach 10 mm in extended length, white in colour with a faint greenish tinge. The slender cerata are numerous and finger-like bearing superficially white pigment specks and internally a green or green-brown digestive lobule. The tip of each ceras is pale. The cerata are arranged in up to 10 rows. The oral and rhinophoral tentacles are smooth and elongated; the propodium is dilated antero-laterally but does not form tentacles.

This inconspicuous aeolid has been recorded from scattered localities all around the British Isles (to 100 m) usually associated with calyptobastic hydroids (e.g. *Sertularia*). Further distribution from Norway and Brittany. There is a record from the White Sea that needs confirmation.

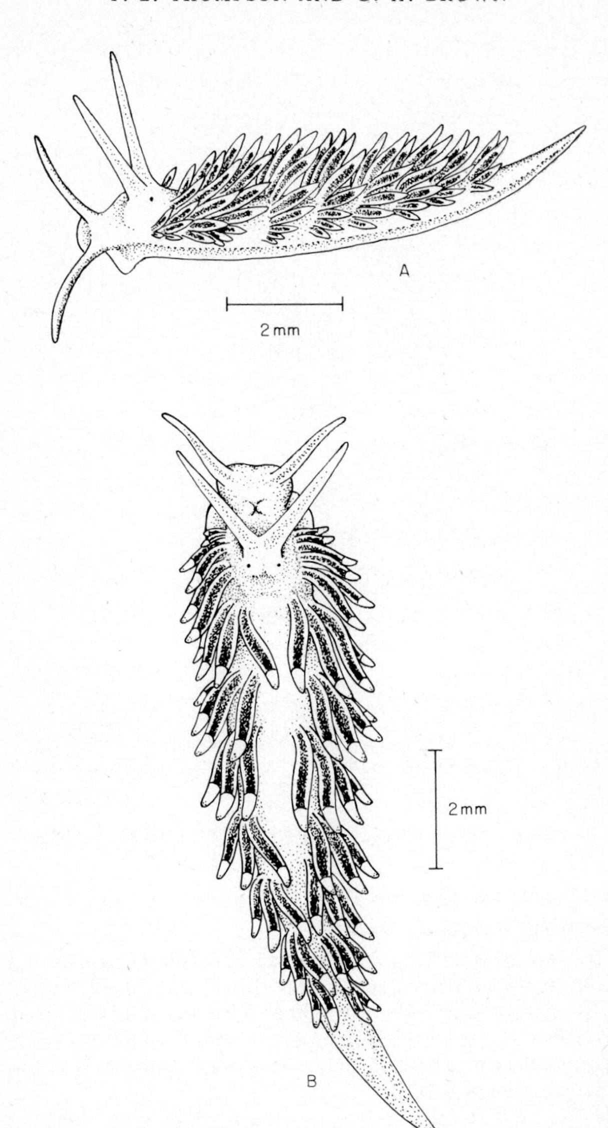

FIG. 102. A, *Trinchesia viridis* dorsal view; B, *Trinchesia concinna* dorsal view (both after Alder and Hancock (1845–55)).

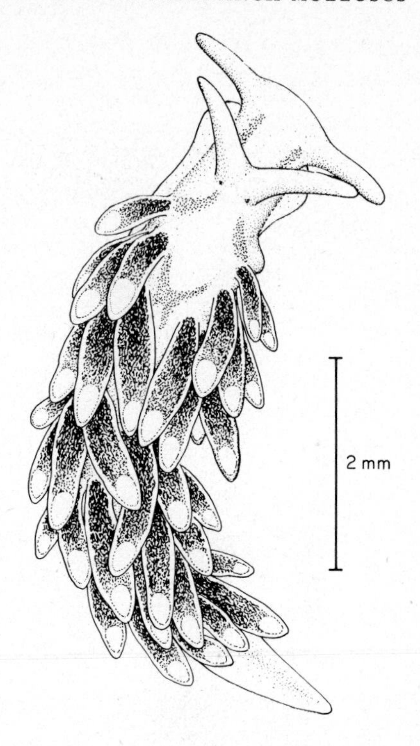

FIG. 103. *Trinchesia foliata* dorsal view (after a colour transparency supplied by H. Lemche).

Trinchesia foliata (Forbes and Goodsir, 1839) (Fig. 103)

Eolida foliata Forbes and Goodsir, 1839

Eolis olivacea Alder and Hancock, 1842

This aeolid does not exceed 11 mm in extended length. The body is stout, yellowish in colour with white superficial spots, and with a very characteristic pink or pale orange streak on either side of the head. The sparse cerata are finger-like with yellow or brown digestive lobules, pale tips and, often, transverse bars or rings of greenish brown. The oral and rhinophoral tentacles are smooth; no propodial tentacles are present.

Trinchesia foliata feeds upon both gymnoblasts (*Tubularia*) and calyptoblasts (e.g. *Obelia*, *Sertularella*) and has been recorded from intertidal and offshore (to 70 m) localities all around the British Isles. Further distribution from Norway, Brittany, to Marseilles and Turkey.

Family FIONIDAE

Genus FIONA Forbes and Hanley, 1851

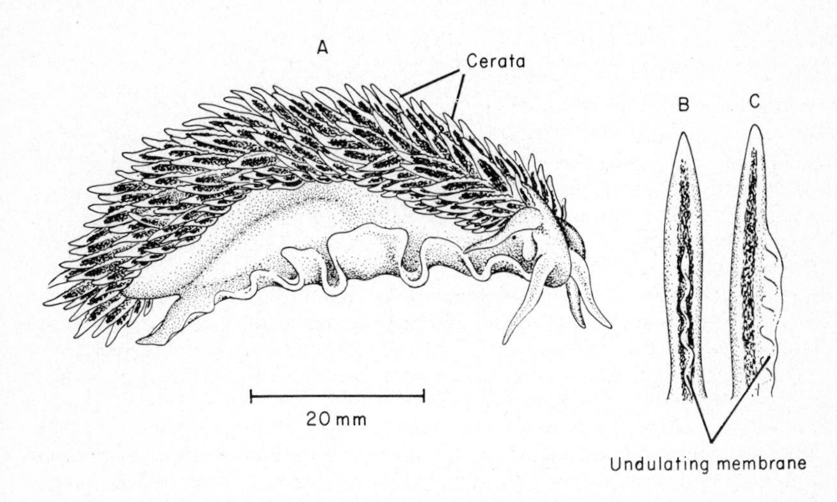

FIG. 104. *Fiona pinnata:* A, whole animal dorso-lateral view; B and C, views of a single ceras (partly after Alder and Hancock (1845–55)).

Fiona pinnata (Eschscholtz, 1831) (Fig. 104A–C)

Limax marinus Forskål, 1775
Eolidia pinnata Eschscholtz, 1831
Oithona nobilis Alder and Hancock, 1851
Fiona atlantica Bergh, 1858

This circum-tropical species has been recorded from British waters only once and that was on the shore at Falmouth in Cornwall more than a century ago. The body may reach a length of 60 mm, pale fawn in colour with scattered opaque white surface flecks. The cerata are finger-like and very numerous containing digestive lobules of which the colour varies with diet. After feeding on the chondrophore *Velella* the digestive gland is blue but after a meal of the barnacle *Lepas* it becomes brown, sometimes very deeply so. The most obviously diagnostic feature of *Fiona* is that the cerata each bear an undulating attached membrane on the mesial face (Fig. 104B and C). The oral and rhinophoral tentacles are smooth. No propodial tentacles are present.

Although *F. pinnata* is always associated with planktonic prey it is not itself capable of active swimming. Outside Britain it has been reported from the Mediterranean Sea, Florida, Australia, Japan and Madagascar, always at the surface of the ocean or stranded with flotsam.

Family CALMIDAE

Genus CALMA Alder and Hancock, 1855

Calma glaucoides (Alder and Hancock, 1854) (Fig. 105)

Eolis glaucoides Alder and Hancock, 1854
Eolis albicans Friele and Hansen, 1876
Forestia mirabilis Trinchese, 1881

The body is dorso-ventrally flattened, reaching 13 mm in maximal length, translucent white or pale cream in colour. The cerata are set in up to 11 pairs of dorso-lateral clusters and each ceras contains a conspicuous digestive diverticulum of which the colour varies (according to diet) from pale yellow to brown. The shape of the cerata also varies so that each ceras may be swollen and pearshaped after a meal but can become slender and finger-like later. In sexually mature individuals the pale yellow gonad lobules can be seen through the skin of the back. The head bears smooth oral and rhinophoral tentacles. The propodium is extremely broad and produced antero-laterally into curved graceful tentacles.

This species feeds almost exclusively upon the eggs of intertidal teleost fish and up to 60 individuals have been found on one batch of such eggs (chiefly *Lepadogaster*, *Liparis*, *Blennius* and *Gobius*). The food is sucked into the alimentary canal and assimilated so completely that the adult *Calma* has no need of an anus and none is present (although certainly present in the veliger larvae of *C. glaucoides*). This species has been reported for scattered localities all around the British Isles and, further afield, from Norway, Brittany and the Mediterranean Sea, to 40 m.

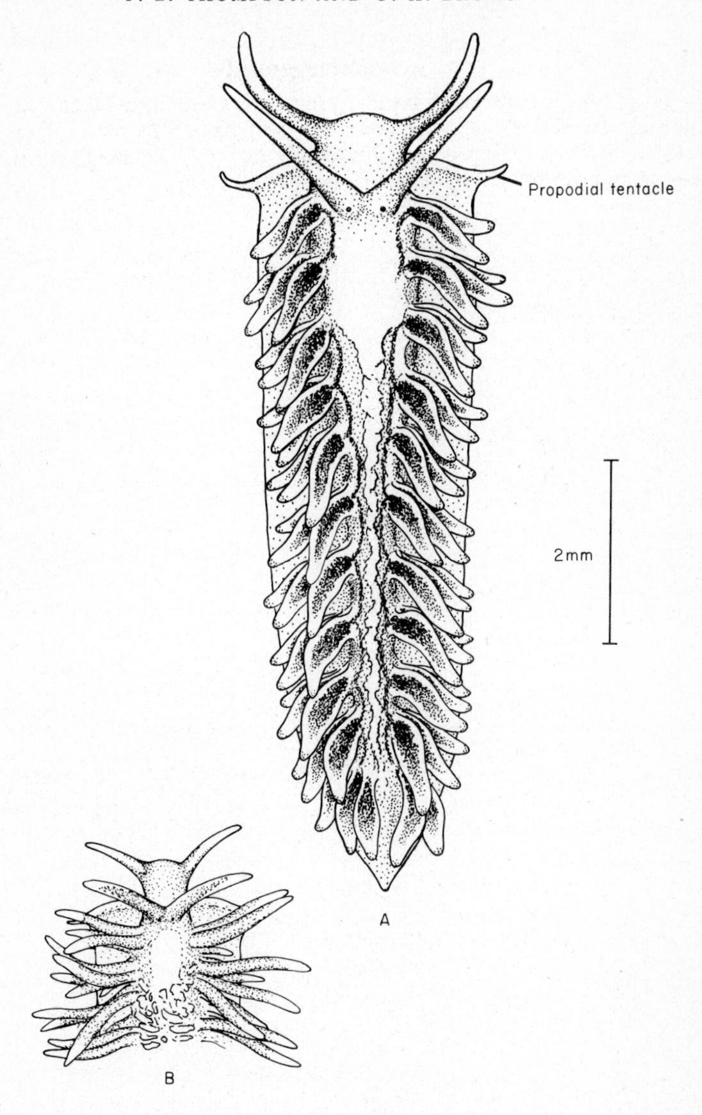

FIG. 105. *Calma glaucoides:* A, whole animal engorged with food dorsal view; B, anterior end of a starved specimen dorsal view.

Acknowledgements

Our sincere thanks are due to the many naturalists who have accompanied us on collecting expeditions over the years, or who have sent material to us. Mr. Bernard Picton has helped us considerably not only by supplying sublittoral material and records but by reading parts of this *Synopsis*.

References

A full bibliography of the British opisthobranchs will form a part of the Ray Society Monograph *Biology of Opisthobranch Molluscs* by T. E. Thompson (to be published shortly). The following references are valuable for the taxonomy of certain difficult groups as well as for a general understanding of recent work on these molluscs.

ALDER, J. and HANCOCK, A. 1845–55. *A monograph of the British nudibranchiate Mollusca.* Ray Society, London.

EDMUNDS, M. and KRESS, A. 1969. On the European species of *Eubranchus* (Mollusca Opisthobranchia). *J. mar. biol. Ass., U.K.* **49,** 879–912.

ELIOT, C. N. E. 1910. *A monograph of the British nudibranchiate Mollusca.* Ray Society, London (supplementary volume).

LEMCHE, H. 1948. Northern and Arctic tectibranch gastropods I. The larval shells II. A revision of the cephalaspid species. *K. dansk Vidensk. Selsk. Skr.*, 1–136.

MILLER, M. 1958. Studies on the nudibranchiate Mollusca of the Isle of Man. *Ph.D. Thesis. University of Liverpool.*

PRUVOT-FOL, A. 1954. *Faune de France 58 Mollusques Opisthobranches.* Lechevalier, Paris.

TARDY, J. 1969. Étude systématique et biologique sur trois espèces d'Aeolidielles des côtes européennes (Gastéropodes Nudibranches). *Bull. Inst. Océanogr.* **68,** 1–40.

THOMPSON, T. E. 1961. The importance of the larval shell in the classification of the Sacoglossa and the Acoela (Gastropoda, Opisthobranchia). *Proc. malac. Soc. Lond.* **34,** 233–238.

THOMPSON, T. E. 1964. Grazing and the life cycles of British nudibranchs. *Brit. ecol. Soc. Symp.* no. **4,** 275–299.

THOMPSON, T. E. 1967. Direct development in a nudibranch, *Cadlina laevis*, with a discussion of developmental processes in Opisthobranchia. *J. mar. biol. Ass. U.K.*, **47,** 1–22.

THOMPSON, T. E. 1972. Eastern Australian Dendronotoidea (Gastropodo: Opisthobranchia) *Zool. J. Linn. Soc.* **51,** 63–77.

THOMPSON, T. E. 1975. Dorid nudibranchs from eastern Australia (Gastropoda: Opisthobranchia). *J. Zool., Lond.* **176,** 477–517.

TURK, S. 1973. *Concordance to the field card for British marine Mollusca.* Conch. Soc. of Great Britain and Ireland, London.

YONGE, C. M. and THOMPSON, T. E. (in press) *Living Marine Molluscs.* Collins, London.

Index to Families and Species

For species and genera the correct names are in *italics*; synonyms are in roman. The page citations in roman are to the text; those in *italics* are to illustrations.